Western Mining

UNIVERSITY OF OKLAHOMA PRESS : *Norman*

Western Mining

AN INFORMAL ACCOUNT OF PRECIOUS-METALS PROSPECTING,
PLACERING, LODE MINING, AND MILLING
ON THE AMERICAN FRONTIER FROM
SPANISH TIMES TO 1893

by Otis E Young, Jr.

WITH THE TECHNICAL ASSISTANCE OF Robert Lenon

Books by OTIS E YOUNG, JR.

The First Military Escort on the Santa Fe Trail (Glendale, 1952)
The West of Philip St. George Cooke (Glendale, 1954)
How They Dug the Gold (Tucson, 1967)
Western Mining (Norman, 1970)
Black Powder and Hand Steel (Norman, 1976)

Library of Congress Catalog Card Number: 76–108800
ISBN: 0–8061–1352–9

8 9 10 11 12 13 14 15 16 17 18 19 20 21

This book is for my mother,
MADGE OLIVER YOUNG,
who has tolerated the vagaries of
two generations of Old West enthusiasts
and who, it seems quite likely,
is in a way to be subjected
to the attentions of a third

Preface

In American frontier studies the techniques of the mining rushes are perhaps the least comprehensible to the student of the era. The most popular chroniclers were not very experienced in prospecting, mining, or milling, and though they could communicate passably well, they tended to gloss over their unfamiliarity with practical matters. The techniques took years to master, even in the frontier period, and all too often the ability to communicate decreased as technical knowledge increased. For example the authoritative works of Egleston, Lakes, and Rickard might as well be written in Old High Gothic for all the good that the unsophisticated reader can derive from them.

It is perhaps inevitable that a profession develops its own special vocabulary, but that can be a mixed blessing, since the vocabulary tends to isolate the profession from the main currents of human experience. Standard glossaries and references are of little utility in deciphering professional books; they shed some light but still leave the problem dimly illuminated. To grasp merely the essentials of frontier mining and metallurgy requires not only a familiarity with the vocabulary but also a knowledge of the rudiments of chemistry and geology, as well as an acquaintance with some very arbitrary mechanics. Perhaps that is why the historiography of the mining rushes emphasizes so heavily what went on in the camps but has so little to say about what was going on underground.

The principal aim of this book is to present by word and illustration a broad outline of the methods by which gold and silver were won from the earth in the Old West. It may additionally serve as an introduction to the literature and as an extended glossary for reference purposes. For these reasons, critical remarks are offered about the utility of the selected bibliography for the historian and student. To assist the reader who wishes to find quickly a definition of a desired term, the italicized page reference in the index is intended to be a guide to the appropriate passage in the text. Some of the phraseology is outdated by present scientific standards, but has been preserved for the sake of reasonable uniformity with quoted passages.

The reader is cautioned that this book should not be regarded as a prospectus for get-rich-quick gold finding. All the high-grade surface bonanzas were located, worked out, and their bones picked clean half a century or more ago. If important mineral deposits remain to be found in North America, they will be prospected by complicated geophysical methods and worked by elaborate milling procedures whose cost is prohibitive for all but very large corporations. Neither should any mining engineer suppose that this book is an attempt to poach upon his grove. A glance will assure him that the quantitative aspects of mining and milling which are his chief concern are covered only in the most general way.

Historians are warned that this account is not an attempt to be either definitive or original. Only the most interesting (to me) strikes, mines, and techniques were selected to point the appropriate moral. The materials consulted were chiefly the best-known memoirists of the period and the major secondary histories. The ideas discussed herein are presented without much criticism of their technical rights and wrongs, for it is a principle of historiography that it was not metaphysical Truth but only what people currently believed to be true that inspired their actions and so provided the raw materials of history.

Barring a few exceptions, the coverage ends with the year 1893. That date marked the repeal of the Sherman Silver Purchase Act—the federal subsidy which had enabled most of the producing mines to stay in business in the face of rising costs. That decade saw the end of the frontier in the sense that Frederick Jackson Turner interpreted it. The period marked the obsolescence of the Heavy Industrial Revolution, whose technical base had rested upon steam power, reciprocating-machine design, and cast-iron or low-alloy steel materials. The ingenious mechanic was giving way to the scientist, and such developments as cyanidation, flotation, and electrochemistry were pushing out the desert-rat prospector and rule-of-thumb millman. Copper was displacing gold as the king of western metals. University-trained engineers were so plentiful as to be available to every working mine, mill, and development. In a word, western mining had ceased to be an adventurous, back-breaking gamble for the Golden Fleece and was becoming instead just another business.

In the preparation of this book I have had the aid and co-operation of a great many people and institutions, all of whom seemed genuinely concerned that knowledge of the old ways should not be allowed to die. I am most grateful to my associate, Robert Lenon, P.E., of Patagonia, Arizona, who practices as a consulting mining engineer and is an amateur, in the best sense of the word, student of history. Bob read, criticized, and furnished many items of great interest and help to the manuscript. In point of fact, his words are woven in and through the narrative, and he deserves full recognition and thanks from me.

The Arizona State University offered substantial aid and encouragement, including a research stipend without which it is difficult to imagine early completion of the project. Various motifs from this study have appeared in print here and there. I wish to express my gratitude to Harwood P. Hinton, Editor of *Arizona and the West*; Doyce B. Nunis, Jr., Editor of the *Southern California Quarterly*; Vivian A. Paladin, Editor of *Montana: the Magazine of Western History*; and the University of Arizona Press, for permission to adapt material originally published in its volume *Reflections of Western Historians*, edited by John A. Carroll. Thanks are due to Donald E. Phillips, former Director of the Arizona Pioneers' Historical Society, for his kind permission to recast a much shorter work published in very limited edition by the society under the title *How They Dug the Gold.* I wish to express appreciation to Don Bower, Editor of *American West,* for similar kindness.

Other individuals who made valuable contributions are registered assayers Archie L. McFarland, of Virginia City, Nevada; Walter Statler, of Humboldt, Arizona; and L. Lee Boyer, of Tempe, Arizona. Academicians who offered great assistance are Odie B. Faulk, Professor of History in Oklahoma State University, Stillwater; Bert Fireman, of the Arizona Historical Foundation; Sidney B. Brinckerhoff, of the Arizona Pioneers' Historical Society, Tucson; Merle C. Nutt, Professor of Metallurgical Engineering in Arizona State University; Watson Parker, of the Wisconsin State University (Oshkosh) history faculty; and Mr. Philip Ross May, of the University of Canterbury, Christchurch, New Zealand. Other contributors are Verne C. McCutchan, State Mine Inspector of Arizona; J. C. "Buck" O'Donnell, of Shaft and Development Machines, Inc., of Salt Lake City, who gave kind permission to reproduce his wonderfully humorous and accurate drawings of the old-timers; Arthur H. Lidman, of Scottsdale, Arizona; Thomas N. McKee, of Paradise Valley, Arizona; and Leonard McCloud, of Chula Vista, California.

OTIS E YOUNG, JR.

Tempe, Arizona

Contents

Illustrations

Plates

Figures

MAP

Western Mining

CHAPTER I

Mineral Lodes and Prospecting Methods

UNTIL NEARLY the turn of the twentieth century, American prospecting reflected no great advancement beyond that of prehistoric ages. Barring the American desert rat's acid test for mineralization,[1] an Egyptian of the Old Kingdom would have recognized instantly what his successor was about. This prospecting, however, was not a science but an art founded upon accumulated lore, a dash of superstition disguised as lore, and a philosophy which is now forgotten beyond recall. Sheer intensity of effort was made the substitute for any sort of theoretical approach; yet, so long as men have been bitten by the Gold Bug, prospecting has never lacked for practitioners. Sooner or later the prospectors of the old school went up every gulch and dry wash in the world. Many, for that matter, are still doing so and are still relying upon techniques and adages which were old when the Great Pyramid was yet undreamed of.

Though there is fairly good knowledge of the prospecting and gold-mining methods of the prehistoric and ancient world, it is next to impossible to reconstruct the ancients' rationale of mineralization. Methods can be observed, but rationales can be learned only in give-and-take discourse; the classical writers, being gentlemen, seldom drank from the same jug with prospectors or miners. Thus writers like Pliny could watch to their hearts' content but usually were denied a view of trade secrets, and so they were at liberty to misinterpret what they saw. The discovery of some uncannily advanced mechanisms developed by the ancients, notably the planetary computer of Antikýthēra and the wet-cell copper-iron batteries of Ctesiphon proves that the ancient technicians were far more sophisticated than the written record indicates, since

[1] This acid test, described subsequently, was itself of respectable age. Nitric acid was known and described in the fourteenth century A.D. by the Arab alchemist Geber (the "Latin Jābīr") and was obtained by the distillation of niter with green or blue vitriol (iron or copper sulphate), a singularly easy reaction to duplicate with crude equipment.

such devices are not so much as hinted at in contemporaneous accounts.

When an aristocrat attempted to exchange confidences with the proletariat, he ran at once into the tradition that prospectors and miners are secretive to a degree, particularly when they suspect that they are on to a good thing.[2] The working classes can scarcely be blamed, for sad experience doubtless quite early taught gold finders to be thrifty with the truth. Herodotus related what must have been a typical example of organized sluice robbing in the placer diggings of India. Broadly hinting that the "ants" in question were an industrious peasantry coyoting along their native stream banks, the Father of History noted that gangs of raiders habitually despoiled the placer men during their rest period, relying upon the speed of their camels to outrun their pursuers.[3] Thus far back—and the tale was old when Herodotus told it—miners and prospectors learned never to tell the general public more than they felt was good for themselves, a tradition still highly regarded in certain quarters.

On the other hand, no one runs off more at the mouth than a miner who is in *borrasca* and wants to unload. Yet even then the whole truth is not to his purpose. He may attend to an embroidery himself or retain a professional publicist, hoping that someone with more money than sense may thereby be persuaded to buy a hole in the ground. Mark Twain related his own experiences at this game in Virginia City:

> They did not care a fig what you said about the property so you said something. Consequently we generally said a word or two to the effect that the "indications" were good, or that the ledge was "six feet wide," or that the rock "resembled the Comstock" (and so it did—but as a general thing the resemblance was not startling enough to knock you down). If the rock was moderately promising, we followed the custom of the country, used strong adjectives and frothed at the mouth as if a very marvel in silver discoveries had transpired. If the mine was a "developed" one, and had no pay ore to show (and of course it hadn't), we praised the tunnel; said it was one of the most infatuating tunnels in the land; driveled and driveled about the tunnel until we ran entirely out of ecstasies—but never said a word about

[2] Leslie Aitchison, *A History of Metals*, II, 375. See also the comment of Thomas A. Jagger, the dean of American vulcanologists, who states that he was early turned away from mining "because of its secrecy and its devotion to profits" (*My Experiments with Volcanoes*, 29).

[3] Herodotus, *The Histories* (trans. by A. D. Godley), III, 102–106.

the rock. We would squander half a column of adulation on the shaft, or a new wire rope, or a dressed pine windlass, or a fascinating force pump, and close with a burst of admiration of the "gentlemanly and efficient Superintendent" of the mine—but never utter a whisper about the rock. And those people were always pleased, always satisfied. Occasionally we patched up and varnished our reputation for discrimination and stern, undeviating accuracy, by giving some old abandoned claim a blast that ought to have made its dry bones rattle—and then somebody would seize it and sell it on the fleeting notoriety thus conferred.[4]

So it is evident that the truth, as prospectors and miners understood it, was extraordinarily hard to come by until the happy day in about the year 1550 when the German physician Georg Bauer (Georgius Agricola) broke down their reserve. Yet it is reasonable to deduce that the early prospector somehow knew the differences between sedimentary and igneous formations and was able to distinguish between the acidic and basic varieties of the latter. A tentatively suggested test would lie in differences of surface texture and color intensity. Sedimentary strata tend toward matte surface and pastel coloration; basic igneous rock is very dark, and its physical forms are highly characteristic. Country rock that affords good prospecting is glossy or vitreous from heat, is light of hue from the silica content, and when mineralized may display quite brilliant shades of color. Undoubtedly some very ingenious theories were worked up to account for mineralization with respect to these differences, but such distinctions would be handed on from master to apprentice as craft secrets. To this day a prospector or assayer will glance at a proffered specimen and classify it as "lively" or "dead," referring to its degree of mineralization. This terminology may be a significant hint about the lost rationale of mineralization. In its absence the encyclopedists had to invent their own theories, which were surely hooted at by the men with mud on their boots.

The archaic prospector must have had a fair idea of not only what mineralized districts looked like but also where such districts might be encountered, since he managed to find every high-grade deposit within reach of classical Mediterranean civilization. At the outset he eliminated the 70 per cent of the earth's surface that is covered with water. That may seem to be a comment on the obvious, but it is not, for suboceanic

[4] Samuel L. Clemens (Mark Twain), *Roughing It*, 308.

prospecting is a great and growing branch of today's hunt for new mineral deposits. Likewise rejected was that major portion of the dry land where surface conditions such as ice or permafrost, deep soil, or dense vegetation permitted only stream-bank placering. What remained was mostly desert country. In such open terrain the earth's crust is well exposed, and much of it can be observed at a glance. As a consequence, most of the chief mineral strikes of the pretechnological period were made in such terrain.

In desert country the sun shines most of the time, and the classical writers concluded that a superabundance of sunshine created mineralization. In a farfetched sense they were right, since elemental gold was assembled in the nuclear furnaces of long-vanished stars. But they were not thinking of that. They merely reasoned *post hoc*, failing to consider that cloudy skies imply rain and that rain brings both vegetation and extreme erosion to conceal the raw surface from the eye. The prospectors themselves probably suspected otherwise but as usual did not volunteer information. Nevertheless, prospecting goes much better on sunny days —cloud cuts by half the ability to detect promising float.

In arid country the experienced prospector would ignore flatlands underlain by horizontally bedded strata of undisturbed sedimentary rock,[5] instead making for rugged regions where subterranean forces had brought up mineralization and exhibited it. The North African desert is a manifest example. West of the Nile the land is flat and little disturbed, and in consequence it was ignored by prospectors until very recently. On the other hand, just east of the Nile runs the Great Rift Valley, and where it crosses the desert at the Sinai Peninsula are the remains of some of the world's oldest copper mines.

Now, of course, the fact is that some varieties of earth disturbance actually militate against any prospects at all. "Basic" vulcanism, which is characterized by quiet, highly liquid eruptive flows which congeal to cinders and sheetlike or ropy black lavas, is one example. Similarly, the

[5] The great industrial minerals of today's world—salt, sulphur, limestone, coal, petroleum, and iron ores—are all of sedimentary deposit. They were of small interest to the world until relatively modern times, and their discovery and exploitation follow rules entirely different from those of traditional mining. When the term mineralization is hereafter used, it can therefore be taken to mean exclusively nonferrous mineralization, with particular reference to gold, silver, mercury, copper, and lead.

related basalts,[6] whose dark-brown-to-black, six-sided columnar crystals are universally familiar (as in the diabasic Hudson River Palisades), are considered the worst of bad hunting. The entire Pacific Ocean basin has only basaltic country rock and vulcanism, and there is no nonferrous mining anywhere in the region.

What is desired is igneous activity on a grand scale, activity which is associated with acidic rock, or rock which is over 50 per cent silica, and hence is classifiable as granitic or one of granite's infinite variations. Such phenomena tend to be associated with profound, sometimes earth-girdling fissuring and faulting of the crust. One such fissure is the Great Rift Valley, which extends nearly from Mount Kilimanjaro to the Gulf of Euboea. Another is the still-active Ring of Fire, which encircles the Pacific Ocean, including among its sectors the Sierra Nevada and the Andes Mountains. Acidic fissuring may carry rich mineralization but may also be marked by catastrophic vulcanism.[7] Even this generaliza-

[6] A basic or basaltic rock has a large proportion of iron and magnesium in its content and possesses less than 50 per cent silica. Diabasic rock is granite with the silica left out, as it were. Both varieties can carry metallic mineralization, but in such extremely low percentages as those of the Lake Superior Keweenawan basalts, whose copper content was probably no more than 0.02 per cent. Such low tenors were undetectable by the earlier prospector and to this day remain unworkable as such. Given appropriate conditions of leaching by warm, salty waters and a receptive stratum such as limestone or conglomerate, however, the mineral content can be removed and redeposited in high concentration. The resulting mineral is native, a very pure sulphide, or a carbonate, all of which are easily prospected and easily worked. Since they are found elsewhere than in the rock of primary origin, unscientific writers often fail to understand their true genesis. A few examples are the Lake Superior native copper and hematite iron and the Ozarkian galena. One critical test of this redeposition is the almost complete absence of the other metals which are found in company with the chief metal in primary deposits.

[7] At temperatures of fusion (1300°F.) basaltic (iron-magnesium) lava is a "thin" liquid whose gas content bubbles off easily and peaceably. Acidic (aluminum-silicon) magma is viscous, retaining its gas content until pressure drops so far that there must be explosive release. The best analogy is the great frothing caused by the uncapping of a frozen carbonated beverage whose gas-carrying contents are more viscous than those of a bottle at room temperature. The "burning cloud" eruption of Pelée which destroyed St. Pierre illustrates the great and sudden generation of such gas. In other cases the acid magma is more slowly evacuated from its underlying chamber in a series of explosions which carry off great quantities of pumice and ash. When the chamber is depleted, a catastrophic collapse produces an implosion characteristic of Krakatoa and Thera and also of Mount Mazama, whose caldera, or collapse crater, now holds Crater Lake. These facts hint that hyperthermal acidic dikes and sills may be deposited well below the surface

tion must be qualified; one of the worst of the killer volcanoes, Pelée of Martinique, is central to the West Indies chain of acidic vulcanism, and yet the region has virtually no mineralization to speak of. In sum, there are rules, and there are exceptions to rules, but the greatest of all rules is that mineralization will probably *not* be in certain places; it *may* be in others; it *will* be only where it is indisputably found, and nowhere else.

Deposit of Ore Values: Fact and Fancy

By American frontier times, conclusions about the genesis of ore existed, like the ladder of Aquinas, on several levels. Greenhorns supposed that such metals as copper and gold were extruded from the earth in solid sills of native metal and that erosion detached grains or nuggets from this "mother lode," whose eventual discoverer would reap the wealth of Croesus. Practical miners shook their heads at such naïve speculation. They too felt that metalliferous lodes had been extruded hot from the earth in a manner akin to squeezing clay out between the fingers, but they knew from experience that the chief mass of such lodes was worthless quartz or other "earthy" constituents. This extrusion theory accorded with unsystematic observation but led inexorably to the conclusion that lodes must become richer with increasing depth, since it was from those depths that their values had originated.

On that premise, the empirical miner and prospector built a number of related fallacies. They felt that the nature of a lode changed for the better as depth increased. They thought that a lode was invariably marked by a "blow-out" outcrop; that the size and extent of this outcrop was a very reliable indicator of the amount and tenor of the ore at depth; and that the lode increased markedly in size below the outcrop. They additionally assumed that atmospheric action tended to leach outcrops of most of their values and that hence ore tenor ought to increase rather dramatically in value with increasing depth. Finally, they assumed that this same atmospheric action "carried down" the values abstracted from the surface, adding them to the already high-tenor ores below.[8] Un-

(perhaps more than three thousand feet below), since otherwise their gas content would foam them into pumice, rhyolite, or tuff. They are ultimately exposed by erosion and lowering of the mean surface.

[8] Cf. G. Montague Butler, *Some Facts About Ore Deposits, passim.*

fortunately, not one of these assumptions is true as a statement of principle, and it was already being theorized by the scientifically educated that far different causes were at work, producing almost diametrically opposed effects. Proving those theories was something else, particularly since the ideas of the "practical" men jibed better with frontier optimism and the requirements of mine promoters than did the growing corpus of scientific opinion.

This "the deeper, the richer" theory was to waste millions in money spent in deep and hence exceptionally costly exploratory pursuit of ore horizons which simply were not there. Sad to relate, most bonanza ore prevailed near the surface where it had been concentrated by the weathering down of "thin" lode values. That was good for the prospector, making discovery easy and sale more profitable, but bad for the mine operator, who hopefully assumed that the ore values persisted or would even increase indefinitely downward. It was also bad for the mining camp which sprang up on unfounded hopes. It led to unwarranted extravagance in everything from milling facilities to local residential architecture, it being poor economy to build for fifty years' operation on ore bodies destined to play out in fewer than ten. At length even the "practical" men were forced to admit that the first days of a mining camp were usually its palmy days. Yet virtually every ghost camp had its handful of resident eccentrics who steadfastly held that the ore body had only been scratched and that wealth in immoderate quantities lay only a bit farther down.

It might be generalized that the period from 1850 to 1890 was a time of confrontation between the empiricists and the scientists, with the public betting on the outcome and usually losing its collective shirt whenever it bet with the empiricists. One could pause here, saying in effect: "This they believed. They were usually wrong and after a while went into *borrasca*. They raised more money, spent it on vain hopes, and then they gave up." Instead, it is better to offer some general outlines of the scientific ideas of ore genesis, if only to illumine history.

Modern geologists say much of magma and consider nonferrous mineralization its gaseous "emanation" product, these emanations rising through the earth as steam rises from a boiling pot. Magma is defined as molten rock at depth, in contrast to lava, which is molten rock upon the surface of the earth. The two are distinguished by the fact that magma contains gaseous products under great pressure, these products being sorted out of lava by the relief of pressure during surficial eruption. There is a growing suspicion that granite and other acidic "igneous" rock are not "true" intrusives, but rather are old sediments which have been penetrated by magmatic emanations—water, silica, metals—melted together by the heat of the adjacent magmatic pocket, and, being relatively buoyant, risen upward in their new granitic phase. This rise may take the form of a balloonlike body melting its way upward as a unit. Should it break surface, acidic vulcanism characterized by great release of gas and the ejection of fine-grained acidic flows such as pumice, rhyolite, tuffs, or even obsidian may result, accompanied by vast explosions. If restrained from erupting, the body may leisurely cool into a hemispherical batholith whose minerals tend to sort and agglomerate themselves into coarse (granite) or very coarse (pegmatite) crystals. In metalliferous formations this sorting ends in a spectrum of onionlike layers, each characterized by a unique if highly "diluted" mineralization shading gradually down into the next. In descending order these metalliferous zones are the silver, the lead, the zinc, the arsenic, the copper, the iron-gold, and the rare metal, including tin or platinum. Since the layers are eroded and removed in the same order, a primary iron-gold ore horizon will probably never give way farther down to, say, silver lead.

Most mountain chains of great size, whether fault-block, folded, or horst, have such central batholiths of granite. At first glance it would seem paradoxical that mountains of tectonic origin should display plutonic features. During the course of their formation, however, their lower elements were subjected to great pressure from the vast overburden, and great heat from being depressed nearly down to the mantle. Their adjacency to the mantle made possible their infiltration by granitizing emanations and their conversion to batholithic form. Erosion then removed the sedimentary overburden; the granitic cores rose by isostasy (a kind of upfloating); and the batholithic domes are ultimately revealed, as is the case of some ranges of the Rocky Mountains (Fig. 1). Somewhat the same processes may have occurred on a larger scale in the formation of the Precambrian continental shields or cores. These enormous reaches of granite are not primary but are the product of tremendous metamorphism of the sediments which originally composed them. Their underlying basement layers are unquestionably *younger*,

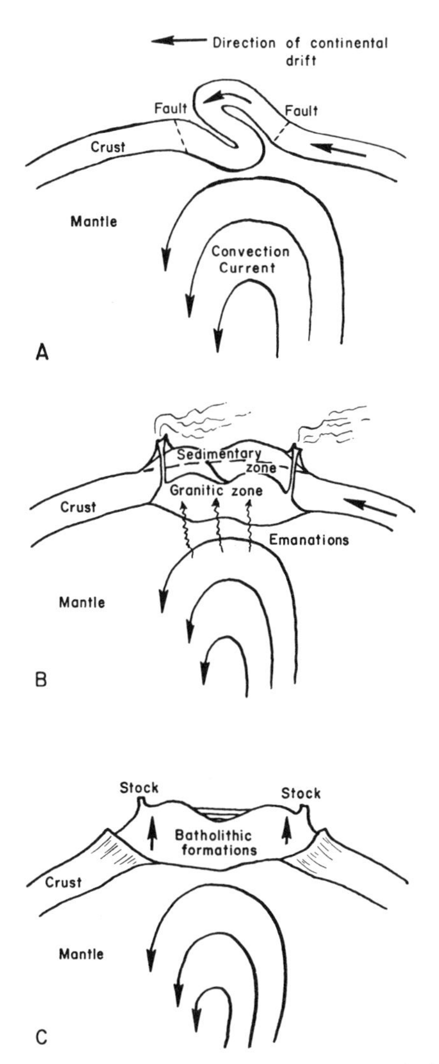

Fig. 1. Formation of batholithic zones in folded mountain building. *A*, initial folding; *B*, lower sediments granitized; *C*, erosion of overburden and isostasy of batholiths.

leading to the disturbing conclusion that whatever originally underlay these Precambrian granites was completely replaced in some manner. Although such processes are very promising for mineralization, they stand as flat exceptions to the geological axiom that whatever lies above is *ipso facto* more youthful than that which it reposes upon.

By whatever means they were formed, these granitic batholiths gradually cooled. As they shrank, they tended to crack extensively. The cracks afforded excellent avenues for the infiltration of water, the latter also being warmed by the residual heat of the body. This water dissolved silica and metallic sulphides[9] from the shells, carried them upward, and concentrated them to form metallic lodes. Not knowing of any part of this process, the frontiersmen were led to think of such lodes as having been forced up molten from depth, extruded through the granite country rock, and then "blown out" through the surface to form the outcrop. Indeed, "blowout" was the term often assigned to an outcrop which was considerably larger in diameter than the lode from which it had originated. Its appearance tended to confirm the false reasoning.

Now and then the magmatic emanations did not spread out to form granites but instead found faults and fractures in the earth's crust and moved up them to form fissure-fault deposits. Again, it was not molten rock but water vapor charged with silica and metallic sulphides which filled these avenues. In such instances the water for some reason tended to deposit only one or two particular metals, suggesting that it had leached them from one "shell" of the magmatic zone. The geologically recent Comstock Lode—so recent that its waters were still scalding-hot—held a silver-lead ore, suggesting that its values originated at the top of a new emanation zone. The lodes at Goldfield, Nevada, consisted primarily of gold in an iron-stained gangue, indicating a somewhat deeper or different genesis. (Peculiarly, tin-platinum ore is almost unknown in North America.) Nevertheless, fond as the geologists were of diagnosing every new bonanza strike as a "true fissure-fault lode," such deposits are actually quite rare.

The two general classes of rock most likely to hold primary nonferrous values are the Precambrian and the Tertiary. Precambrian is a catch-all term embracing all formations older than 520 million years: it comfort-

[9] To maintain uniformity with the sources, *sulphur* and its variations are rendered in the nineteenth-century spelling.

ably includes 3.5 billion years of terrestrial history, or seven times as much as all other periods together. Precambrian deposits tend to be igneous or so metamorphosed as to be treatable as igneous, and almost everywhere but in Canada and Scandinavia they are overlain by enormous layers of sediments or far more recent volcanic flows. They do seem, however, to be the "cores," or heartlands, of the several continents, so that no matter how the continents may wander about the globe, collide with each other, or go careening off at tangents, their Precambrian cores remain fairly intact and in place relative to the outline of the continent as a whole. In addition to their stability, Precambrian formations seem to be very profitably mineralized. There is no magic about this fact; these archaic layers have been in place for so long that the cumulative chance that they will contain something worthwhile has been proportionally increased.

The other period of geology interesting to the mineralogist is the Tertiary (7.5 to 2.0 million years past), characterized by an *apparent* renewal of plutonic activity that had seemingly been suspended during the Paleozoic and Mesozoic eras of intervening time. (Twenty years ago the preceding statement might have been a flat assertion; today geologists are inclined to be less certain.) In the Tertiary is found the present phase of continental drift associated with vulcanism and mountain building of both the folded and the tilted-block varieties. Indeed, some of the Tertiary mountains such as the Alps and the Himalayas are so youthful that their soft sedimentary overburdens have not yet been sufficiently eroded to reveal their batholithic cores, which are still probably in process of formation some ten miles below sea level. The frontier empiricists were on better ground in this regard; they soon learned to confine their attentions to Precambrian and Tertiary deposits, disregarding the vast intermediate layers of sediments which were, generally speaking, of no interest to them.

Now and then Tertiary formations are commingled directly with those of Precambrian times, and joy reigns supreme. The presumption is that the recent activity heaved up and exposed the old basement rock, which would otherwise have remained buried out of sight. In these regions of commingling (as in the Bradshaw Mountains of Arizona or the Black Hills of South Dakota), relatively new mineral lodes may be found cheek by jowl with some extremely old deposits. In Arizona the great Tertiary copper stocks are found relatively near the Precambrian United Verde–U. V. X. (United Verde Extension) copper lode, although all they really have in common is mere physical proximity.

The Great Rift Tertiary fissure was instrumental in the dawn of the Bronze Age in the eastern Mediterranean world, providing that area with its first school of practical prospecting and metallurgy. The gold of Punt, avariciously mentioned by the Egyptian kings, was both placered and mined from lodes on either side of the Red Sea, which occupies this segment of the Rift. The copper mines of Sinai, where the Rift divides into two arms, were well known centuries ago. The northeastern arm of the Rift extends into Palestine, forming the Dead Sea depression, a fact which has given rise to speculation about the fate of Sodom and Gomorrah (the Old Testament description of their destruction sounds remarkably similar to other accounts of catastrophic "burning cloud" acidic eruptions). At the head of the northeastern arm lies the island of Cyprus, whose very name is the etymological root of the word "copper." The Phoenicians had their headquarters along this region of the coast, and they are generally regarded as the salesmen, if not the originators, of classical metallurgical technology.

The northwestern arm of the Great Rift is submarine, its course passing near Crete—a place of incessant earthquakes—probably right under the island of Santorin (Thera), and thence north to Attica, Macedonia, and through present Yugoslavia. It is instructive to note that Santorin (which some think was Plato's Atlantis) blew half of itself away in an appalling acidic blast during the late Bronze Age, that the silver mines of Laurium played a critical part in the economic history of Periclean Athens, that a substantial gold lode financed the administration of Philip II of Macedon, and that mercury ore is found still farther north in Yugoslavian Idrija. The collation of these geographic facts does not appear to have occupied the attention of the classical authorities, but it seems that the ancient prospectors and miners wasted little time in heading for the right areas.

The Great Rift is a graben formation, the effect of two parallel faults with the crust between subsiding to form a deep trench. The opposite effect, where the land between two parallel faults is elevated rather than depressed, is called a horst; the highly mineralized Harz Mountains of Saxony are a good example. On the western margin of the United States the Sierra Nevada Mountains are of tilted-block origins, with the primary fissuring to the east, where their more abrupt faces lie. It is suggestive to reflect that a line of once-great mining camps—Virginia City, Tonopah, Goldfield, and Searchlight—parallels this plane of crustal weakness. The Colorado Plateau has likewise an abrupt rim to the southwest, and a few miles beyond lie the Bradshaw Mountains of Arizona, with their gold mines; some hundred or more miles farther south are the enormous low-grade copper stocks of the Sonoran desert. Clearly, acidic activity and fissuring constitute a siren call to the prospector (Fig. 2).

The chief primary (that is, original, igneous) gangue or vein rock associated with metallic mineralization is silicon dioxide, found in its purest form as quartz. Silicon dioxide acts as a "trap" to catch nonferrous metal, and the greater the silicon content of a rock the more ef-

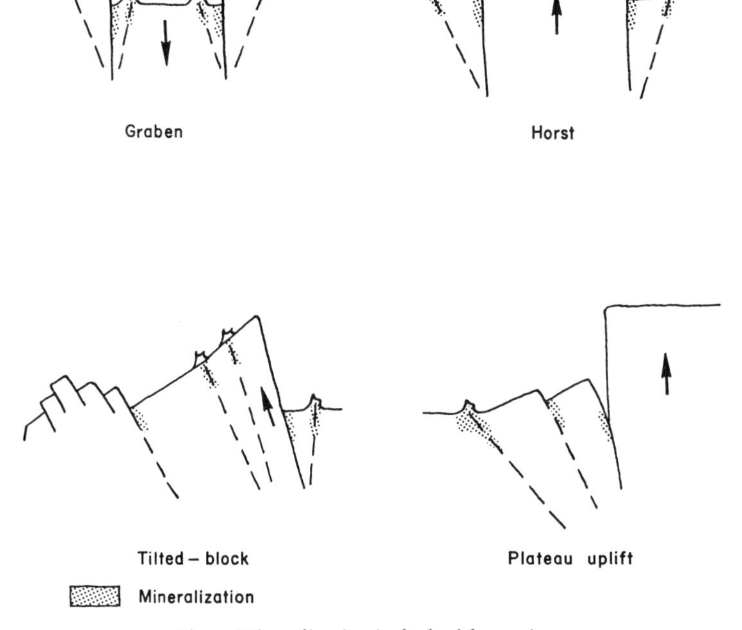

Fig. 2. Mineralization in faulted formations.

fectively it traps mineralization. The vitreous, cryptocrystalline form of quartz is nearly 100 per cent silicon dioxide; granite has, by definition, more than 50 per cent silicon dioxide; porphyry and rhyolite—noncrystalline quick-hardened rocks of somewhat the same composition as granite—can carry substantial values under special conditions. The ancient writers, such as Diodorus and Pliny, were aware that bonanza gold lodes were found in a "white and shining rock" but thought that the rock was marble (metamorphosed and somewhat dehydrated calcium carbonate). Did the miners and prospectors know the difference between quartz and marble? It is difficult to believe that they did not, for quartz and marble have differing density and structure, they react differently to impact and to weak acids, and they occur in different formations and regions. That the same mistake was made repeatedly by the writers is therefore a clue to the inadequacy of their field research, as well as to their ignorance of the mineralization theory of their contemporaries.

Primary ore deposits take three general forms, of which the greatest is the enormous core or stock, often miles in diameter, of granitic rock which carries very low grade (0.5 to 2.0 per cent) sulphide (pyritic) mineralization. Although the mineral is disseminated through the stock with fair uniformity, it tends to be slightly more concentrated in the center than at the periphery. These dissemination structures were ignored as such until the twentieth century. The ancients could make nothing of their values, although today they form the reason for being of the entire western United States copper industry. The second form of primary ore deposit is the fissure created or filled by igneous intrusion. These intrusions tend to follow pre-existing faults or to work their way between the bedding joints of the strata. Fissure-deposit mineral percentages tend to be much higher than those of dissemination structures. They are also easier to mill, since the fissure lode may be further enriched by secondary water action sometimes called supergene. The third form of primary ore deposit is the replacement deposit, in which the mineral values have migrated from the original acidic rock to replace or to be deposited upon the contact zone of a "susceptible" rock, usually and preferably limestone or its chemical analogue, calcite.[10] The latter tend to be "carbonate" or "chloride" deposits, and they are very readily smelted in crude furnaces.

[10] John S. Shelton, *Geology Illustrated*, 366.

Quartz or high-silicon rock can also entrain metallic sulphides which are working their way upward following fissures and vents. Water, present either as superheated liquid or as steam, dissolves low-lying metallic salts, carries them up the growing lode, then redeposits them in cavities between the sides of faults, within the bedding joints of strata, or wherever else heat and pressure suddenly drop off. This phenomenon, for long a theoretical postulate of modern mineralization theory, has lately received some confirmation:

> Unexpected support for this interpretation has recently come from a mile-deep well drilled next to the Salton Sea. It was located near some active hot springs, and recent volcanoes with the purpose of tapping natural steam for power. After a few months . . . the well had reached briny waters with temperatures well above 500° F.—the highest ever encountered in a well. Still more remarkably, after flowing through a large 275-foot-long horizontal pipe for three months these solutions had coated the inside of the pipe with an estimated 5 to 8 tons of dark deposit containing about 20% copper and lesser but astonishingly large amounts of silver and a host of other unusual elements normally found concentrated only in ore deposits. It would appear that here, for the first time, man may be able to study ore-depositing solutions at work instead of drawing inferences millions of years after they have cooled off or dried up and their host rocks have been exposed by erosion.
>
> The question of why a few intrusives are accompanied by active mineralization, while the vast majority seem not to be, remains a mystery.[11]

The same heat and pressure are simultaneously metamorphosing the country rock in contact with the lode, and perhaps playing other tricks as well. For instance, the original mineral content of a primary deposit can be removed by ground water and concentrated in the contact zone of the adjacent wall rock. Thus the primary lode is relatively barren, but the once-unmineralized country-rock casing has become a very rich secondary deposit (Fig. 3). For all these reasons, the propector never cared for consistent uniformity in the country rock, irrespective of its nature, but preferred confused, commingled displays, featuring, it was hoped, a multitude of quartz threadlets and stringers penetrating heavily metamorphosed formations. The miner got a bonus from this phenomenon, for high-silica veins are often quite distinct in color and texture

[11] *Ibid.*, 367.

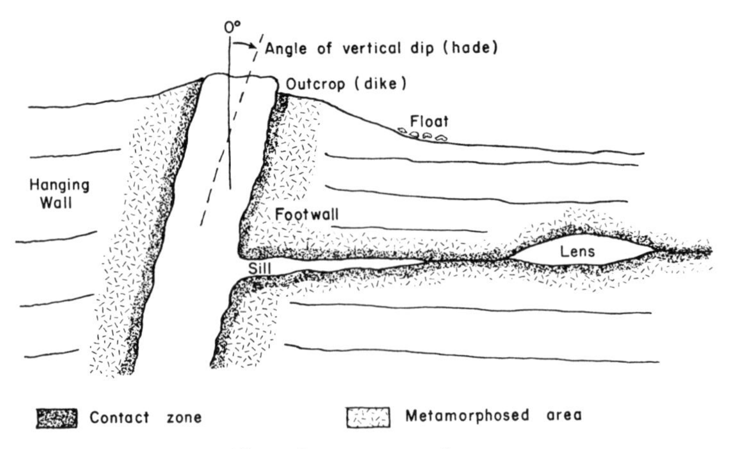

Fig. 3. Outcrop nomenclature.

from the wall rock and so serve as handy guides to lode and contact zone to the man who follows them into the earth.

Prospecting in the West

Having wandered into promising terrain—often following up rumors of casual finds of native metal by the aborigines—the prospector searched for specific indications of mineralization, going partly on traditional signposts and partly on intuitive, perhaps even artistic, judgment. Highly mineralized country is quite difficult to describe but is unmistakable when once encountered. Quartz seems to be everywhere—in Gold Canyon, the gateway to the Comstock Lode, there is hardly a rock of any sort which is not penetrated by these white, vitreous threadlets. The very hills may be solid quartz—a few miles south of Tonopah, Nevada, Mount Mizpah stands alone and conspicuous among the dreary basalt ranges, shining white from afar, a beacon as it were to the mineralization of Tonopah and adjacent Goldfield.[12] The country rock itself may have a rainbow diversity of color, the result of chemical changes through metamorphic activity, and its structure is distorted, even appearing

[12] A few miles northeast of Scottsdale, Arizona, stands the similar Quartz Mountain, a cone of quartz so pure that it appears snow-capped. Unfortunately, it has no value save as a source of roofing gravel. Frontier miners referred to such barren intrusives as bull quartz.

"burned" from the iron content. The impression is one of a riot of shades reminiscent of a Van Gogh painting.

Even in brushy terrain, surface anomalies were the first to be investigated, since they hinted that the rock of which they were composed was either harder or softer than the country rock and hence probably of differing composition. When harder, it might protrude as a white, maroon, or rusty-brown outcrop, whose sides would emerge at a peculiar angle. When softer than the country rock, erosion would have produced a series of depressed anomalies, systematized as

1. A trench or ditch that does not run directly down the slope of the hill or mountain.
2. A sudden change of slope.
3. A sharp notch that crosses a ridge that has a rather uniform altitude on both sides of the notch (hinting at a fault at this point).
4. Several springs of water in a line with each other.
5. A sudden change in kind or quantity of vegetation.
6. A change in the nature of the rock fragments.[13]

Finally, any dike or sill not clearly of basaltic origin called for investigation, as did the exposed walls of dry gulches or arroyos, where the formations have been obligingly exposed for survey at a glance.

Illogically enough, sedimentary formations (though laid down for the most part on the beds of ancient seas and thus seemingly as far removed from igneous action as water is from fire) were not without their attractiveness to the prospector. If an igneous intrusive proved barren, the contact zone of metamorphosed sediments, such as schist, shale, or slate, might well bear scrutiny. Old sandstones and conglomerates might be "fossil" placers, enriched from long-vanished lodes but protected from further erosion by submarine sediments or by lava cappings. They could be traced only by float, since they were usually devoid of any blossom. Gold in particular is so chemically inert as to be imperishable to a fantastic degree. Gold bullion salvaged from the open sea is as untarnished as though it had been lost yesterday instead of centuries ago. Thus free placer gold can turn up in virtually any sort of formation, no matter how theoretically unpromising, provided only that action of some sort at one time deposited it there.

[13] E. D. Wilson, J. B. Cunningham, and G. M. Butler, "Arizona Lode Gold Mines and Gold Mining," University of Arizona *Bulletin*, Vol. V, No. 6, 248–51.

The prospector was particularly interested in color, and his heart beat faster when he found the blossom of mineral salts "painted" upon a ledge or outcrop. Such blossom was usually of deep, primary shades: the brown-to-yellow of iron, the green or blue of copper, the black of manganese, the light yellow of molybdenum or lead, and the lilac of cobalt. The color not only indicated mineralization but was usually a good clue to the values in the lode. The rusty brown of hematite (iron oxide) was regarded as a signpost of gold, for gold is quite often associated at depth with the iron pyrites which weather down into hematite. The Grand Central lode at Tombstone, covered entirely by scree, was nevertheless traced by the deep-brown fan of hematite stain which had spread downward from its outcropping. Copper chlorides and the carbonates' blue and green hint at silver: Panamint was discovered by contemplation of a cliff glowing with their peacock shades.[14] Indeed, almost any intense color *but* lemon yellow or metallic yellow constituted a call for investigation.

For the most part, lemon yellow was but an indication of the laboratory metal uranium. If the color had a metallic glint, it was almost certainly copper pyrite (chalcopyrite) or iron pyrite, the fool's gold of common lore, unless it was the even more worthless scales of yellow mica. Eye, hand, and teeth could instantly distinguish between the false and the real article. Pyrite "blinks"—that is, if the sample is rotated slightly in the light, its plane crystal surfaces wink, or change their reflective pattern. Gold specks are rounded and have a steady, less brassy shimmer. If a crumb of the substance is placed on a hard, flat surface and tapped lightly with a hammer or knife handle, it will flatten if gold, but will shatter into dust if pyrite. A bit of the dust taken between the teeth feels distinctly gritty if pyrite; gold feels softer and produces no gritty sensation. Still and all, the very presence of pyrite could indicate some degree of mineralization, whereas mica is no particular indication at all.

If the prospector discerned nothing tangible in a visual survey of otherwise promising country, he was by no means at a loss. Feeling in his bones that something worthwhile ought to be about, he would hunt up a camp site, make himself and his burro comfortable, and then lay

[14] Neill C. Wilson, *Silver Stampede: The Career of Death Valley's Hell-Camp, Old Panamint*, 30 (hereafter cited as *Silver Stampede*); Arthur Lakes, *Prospecting for Gold and Silver in North America*, 242 (hereafter cited as *Prospecting*).

out his tools in preparation for a search for "float." This term embraces mineral particles and scales eroded from a lode and carried away by water action. By finding float and tracing it back to its point of origin, the prospector might well discover the lode where it had originated. In dealing with float (from one viewpoint, all placering was nothing *but* dealing with float), one either found it by inspection—rusty-looking, "spongy" quartz in outwash gravel was hailed with delight—or sought it by digging. When it was found, its physical condition was significant. Round, smooth float hinted at much abrasion by wind and water, and hence a long journey from the point of origin. If the float was rough and angular, there had been little abrasion, and hence the source might not be far from the point of discovery. Two pieces of similar mineral were far better than one, for an infinity of causes might account for one being where it was, but two or more pieces argued that a straight-line sequence of detachment, travel, and discovery was involved. The obvious next step was to estimate the direction and distance which that travel had involved.

Much has been made about the accidental discovery of rich lodes by animals. The stories take two general forms. The first tells about an animal uncovering ore while pawing the ground or grazing. The other tells about a man picking up a rock to throw at a recalcitrant stray and then feeling its weight or seeing it broken open to reveal intriguing contents. There is little to such stories. Of the few tales which survive close examination, some common points appear. Almost without exception the incidents occurred in early morning at a prospector's or traveler's camp site, seldom or never during the course of travel in the open or while making camp at dusk. Since a camp site is chosen with an eye to vegetation, water, and shelter—that is, close to an abrupt elevation and adjacent to a spring or running creek—in well-mineralized country that is just the sort of place to go prospecting, particularly since spring water emerges from geological discontinuities. Thus, if the animal's antics had not resulted in discovery, a systematic search certainly would have.

If float was sought with deliberation, the prospector commenced operation at the head of a promising wash or in the throat where a creek emerged from the heights. At that point heavy minerals would be nearer the surface and water more frequently obtainable, both because bedrock is, by definition, closer to the surface there and because heavy float tends to dig itself in the farther it travels downstream. The same is true of water. With pick and shovel the prospector dug his prospect pit down to the bedrock, saving soil specimens from the lower layers. If no water was available he subjected his gleanings to dry washing; that is, he spread out a sheet, then gently plucked and puffed at the sample laid upon it until the lighter constituents had been blown away, leaving only the heaviest fragments behind. If water was handy, he was content to pan an occasional shovelful of dirt, handplucking pebbles and washing away the lighter soil over the edge with circular, gently flipping movements until perhaps half a spoonful of residue was left. Either way, the prospector was most often rewarded with nothing. Now and then, however, he would be encouraged by a procession of golden specks (which might also be pyrite) or the black granules of silver lead (which might also be galena or spicular hematite) in the drag of his pan.[15]

It occasionally happened that the prospector had reason to believe that gold was present but that his pan was deceiving him. The prospect pan was frequently used for cooking or eating, and it might be greasy. In that condition it was difficult to retain very fine flour or flood gold dust in the drag. When the presence of flour gold was suspected, the remedy was to build a fire and heat the pan red-hot to burn off any grease.[16] As long as the pan was red-hot, any suspicious-looking dark float could be sprinkled over it. If a puff of white vapor arose, a telluride mineral was present. Many metals form tellurides, but it occurs so frequently with gold that a positive reaction was a very promising indicator (Fig. 4).

Panning and dry washing are among the most primitive prospecting and placering methods; yet they have never died out. Whenever there is severe economic depression, virtually every halfway-promising wash and arroyo in the southwestern deserts is occupied by some poverty-stricken soul attempting to garner a few grains of gold with a mechanical descendant of the "Hungarian" dry washer, specimens of which still survive.

This dry washer was a cheap, crank-operated device consisting of a

[15] The drag is the last half spoonful of heavy residues left after washing. The term is probably a corruption of *dregs*. The tail is the cometlike line of gold specks left after a successful washing.

[16] Richard B. Hughes, *Pioneer Years in the Black Hills* (ed. by Agnes Wright Spring), 29.

Fig. 4. Panning (pit in background). From Thomas Egleston, *The Metallurgy of Silver, Gold and Mercury in the United States* (1887–90).

pair of slanted screens, a canvas-bottomed riffle box, and a built-in bellows. When a shovelful of very dry dirt was placed on the screens and the crank turned by hand, the screens moved against each other, automatically passing the finer material and rejecting the coarser pebbles, which rattled down the surface and away. The sand was guided by a canvas apron into the tilted riffle box, where the bellows subjected it to puffs of air coming up through the canvas bottom. Lighter dust was blown up and away, leaving the heavier particles trapped behind the riffle bars.[17] At the end of a hot and miserable day's work, the operator collected his gleanings, washed them in a small pan or horn spoon, and hoped to find fifty cents' worth of gold dust—wretched pay even for a Chinese in the days of '49, but sufficient for beans and a bag of Bull Durham in Hooverville.

The nature of the residue on the sheet or in the prospect pan went far to determine the prospector's next course of action. Granules of hematite or crystals of pink garnet were strongly indicative of gold prospects, while free gold explained itself. Tiny black cubes were likely to be

[17] F. J. H. Merrill, "Dry Placers of Northern Sonora," *Mining and Scientific Press*, Vol. XCVII (September 12, 1908), 360–61.

galena—lead sulphide. If form or color gave no decisive clue to the nature of mineral particles heavy enough to be panned out, it was necessary to determine whether they were indeed metallic and, if so, what metal or metals were present. The prospector collected sufficient particles to fill a spoon, dried them, and scraped them into a small cavity cut in a charcoal block. He covered the sample with a layer of bicarbonate of soda, then used a blowpipe and candle flame to smelt the batch. (Somewhat the same effect could be approximated by burning the sample with damp gunpowder in a closed container or just by edging it into close proximity to a campfire's coals.) This chemically reduced the mineral by driving off sulphur, chlorine, or carbonate, leaving a metallic bead or button behind. If the specimen was ferrous (iron, cobalt, nickel), whose reduction temperatures were much too high to respond to this treatment, it merely turned brown or red. Copper would announce itself, however, by the color both of the mineral and of the smelted bead. Black gold (gold telluride) would reduce to a characteristic yellow pellet. A white metal bead could be rolled in slightly dampened salt, then exposed to the sun to show black streaks symptomatic of silver. If the button could be easily remelted, it probably had a great deal of lead in its make-up. Tin could be ignored, since it practically never occurred in the West. In short, no matter how crude his equipment, a wily investigator could force the rocks to give up at least some of their secrets.[18]

Supplementing or even replacing the archaic blowpipe test was the acid test, which would rapidly distinguish silver and gold from the white base metals, as well as identify lead and copper promptly. It is interesting to note that early gold prospectors first roundly excoriated the silver of the Comstock Lode as an aggravating "black sand" which clogged their riffle boxes. Later the pale dust made up such a major proportion of the free gold dust that it was called bogus gold. But those particular prospectors knew so little of field assay work that none thought to apply the blowpipe or acid tests to see exactly what it was they had encountered.

[18] But by no means all of them. Arsenic, antimony, tin, and zinc were more difficult to distinguish because of the ease with which they alloyed themselves with lead or silver. The ancients knew only the seven traditional metals (gold, silver, copper, mercury, lead, tin, and iron), failing to identify arsenic and zinc as such, probably because of the above difficulty. Even today the base white metals give trouble in field assaying.

Some chemistry necessarily enters the picture at this point. Gold is usually found in the pure metallic form as free-milling gold (the tellurides and pyritic exceptions will be discussed later). Silver, on the other hand, is naturally and normally found below ground in the form of sulphurets, or silver glance. Where such a lode crops out into the air, the atmosphere goes to work on it, gradually converting the sulphurets to chlorides or, occasionally, carbonates. These silver chlorides are detached by erosion, tumble away, and constitute black granules of float. The prospector usually goes for the lode or vein, but in the Old West some men made a living of sorts scavenging the washes below the lodes for silver chlorides and the dumps below the mills for sulphurets, which they exhumed and sold. These people were called chloriders, which was not exactly a complimentary term.

Chloriders at work were well described by William Wright (the Washoe chronicler who wrote under the name Dan DeQuille). He noted the activity of two Mexicans who had set up shop in Gold Canyon, the gulch running south from the Comstock Lode. The first Comstock silver mills had been erected up the same canyon, and their tailing was often dumped down its course. Said DeQuille of the chloriders:

> They caught the tailings in a small reservoir, from which they took them and spread them on a table [planilla] that stood at an inclination of about thirty degrees. They then threw water over the tailings with a small dipper, beginning at the top of the table and gradually working downward until they reached the bottom, at which point, where the end of the table rested on the ground, would be found some pounds of sulphuret of silver, with some particles of amalgam and quicksilver that had escaped from the mills . . . . [They] worked all summer, and the supposition was that they were about "making grub," but after they left, the butcher of whom they obtained their meat stated that they took away with them about three thousand dollars each; that they were in the habit of bringing their bullion to his shop every Saturday night to weigh it.[19]

Meanwhile, the prospector is staring into his pan. Should the drag display enough of the processional speckles of gold, and should enough

[19] William Wright (Dan DeQuille, pseud.), *The Big Bonanza: An Authentic Account of the Discovery, History, and Working of the World-renowned Comstock Lode of Nevada* (New York, 1964), 94–95 (hereafter cited as *The Big Bonanza*).

water be present to make placering worthwhile, the search was frequently terminated right there. But should the color be sparse, be covered with the ferric "rust" or black-manganese film that inhibited efficient placering, show physical indications that it had come from a nearby source, or be mixed with other heavy minerals, it might be best to abandon the prospect hole and go looking for the lode. Size and shape of the color were determined by scrutinizing the drags of several panfuls, concentrated in the horn spoon or in its replacement, a small iron skillet sawed in half. Largish, rough dust or square crystals hinted that the source was near at hand; flattened and "frayed" grains, or flourlike lenticular dust proclaimed much peening on stream bottoms, hence a long trip down and a prolonged search.

Should the lode of origin not be immediately apparent, or should the country not afford a good hint to lines of wash, the prospector employed statistical sampling methods, invented by his profession thousands of years before mathematicians thought of them. He began to dig like a gopher, taking dozens of scattered soil samples from as many upstream prospect holes, and carefully washing or winnowing portions from each. After great labor, his mentally tabulated results might show that the holes within a given area held mineral float and that the holes on either side of the area did not. More digging amplified these findings, so that at the end he could lay out an imaginary triangle in three dimensions, whose apex ought to lie on the point of origin, even though it was concealed by brush or talus.[20]

A side development of this art was practiced in California gold-rush times by the pocket miner, who worked the banks of mountain streams in search of point deposits, or pockets, of gold. He ambled along the bank, digging and washing at intervals until he found a grain or two of gold in the drag of his pan. He then camped and began to take more samples, working back and forth along the water's edge until he had a firm idea of the extent and depth of the color's distribution through the soil. Having laid out the base of his triangle, he then began ascending

[20] Costeaning, the digging of a prospecting ditch, is an example of this sort of work. A strong hint is afforded by this technique that the legendary concealing of a rich lode, either by covering it with rock or by shooting down the adit, would not deceive for ten minutes any prospector of experience. Given the slightest indication of the inevitable downstream float, he would head for the "lost mine" like a bear to a beetree.

the bank, digging line after line of prospect holes. If he was fortunate, he would presently find the area of color narrowing, deepening, and concentrating. When he had established firm vectors of location and depth, he could go directly to the point indicated and find what he might find: "a spadeful of earth and quartz that is all lovely with solid lumps and leaves and sprays of gold."[21]

As the desert prospector worked his way upward, the samples would likewise narrow the area of search; and as he approached the lode, the float would cease to be free, but would instead be held in bits of vein rock. Given luck, there it would be at the apex of search. But what *would* it be? Most often, a greater or lesser ledge of perfectly conventional ore, none of which would appear at all attractive to the eye of the inexperienced unless, as occasionally happened, water and weathering had concentrated lumps of pure native metal at the surface. On the other hand, some of the best gold-bearing quartz is "rusty," imparting the impression of having served as a doorstop in a junk yard. Silver quartz is a dingy gray, varying from oyster to black. "Fossil" gold-bearing sediments and conglomerates give no visible indication whatsoever of mineralization. Now and then, however, a strike would be something completely astounding and contrary to nature (as prospectors understood nature). In Arizona Territory in 1863 the Peeples party found a graveling of very substantial gold nuggets exposed on the flat top of Rich Hill Mesa (sometimes inelegantly known as the Potato Patch).[22] Now as a rule nuggets are found next to the bedrock of watercourses, or at any rate declivities, not roosting about on the tops of mesas. Nevertheless, there they were, and the Peeples party made off with the lot, scarcely pausing to give thanks. There was, of course, a reasonable explanation, but the odds of the phenomenon being repeated are very, very slender.

Having found an apparently mineralized lode of some description, the prospector proceeded to knock off a chip, apply his wetted tongue to the raw surface to remove dust and highlight the contents, and peer long and intently (in later days, with the help of a pocket magnifying glass) at the indications revealed: specks of yellow gold or a thin blue line of sulphurets. Indeed, the occupational stigmata of the prospector

[21] Clemens, *Roughing It*, 436–38.

[22] Charles B. Dunning and Edward H. Peplow, Jr., *Rock to Riches: The Story of American Mining*, 61 (hereafter cited as *Rock to Riches*).

might well have been a calloused tongue, since this organ was an important part of his armamentarium. When prospectors met, they did not exchange the Kiss of Peace but swapped rock specimens, to which each clapped his tongue and at which each peered before exchanging congratulations (modified by mental reservations).

If the saliva test augured well, the prospector set in train a rough field assay for the purpose of determining whether the ore would be worth working, always beginning with an egg-sized sample which experience showed would weigh so near two ounces as to make no difference. (Prospectors seldom encumbered themselves with balances and weights; if a weight was necessary, water was measured out in the one-fluid-ounce glasses which were readily obtainable in any settlement large enough to support a saloon. A thin batten of hardwood, marked like a steelyard for the weight positions, was set upon the edge of a rock to act as a rough-and-ready pan balance.) For a determination of gold content alone—as with a conglomerate ledge—the sample was pulverized in small iron mortar, then boiled vigorously with mercury, salt, and soda. The liquid was poured off and the mercury amalgam washed free of the earthy residues. The amalgam pill was wrapped in a twist of paper or parchment, roasted upon an iron shovel to drive off the mercury, and the bit of sponge gold remaining estimated with a knife gauge.

To make a general determination, the investigator likewise began by pulverizing the ore sample. He panned off the gangue, collected the drag, and dried it. If the only acid he had with him was hydrochloric acid, he then had to smelt the sample with his blowpipe before dissolving it; if he had nitric acid, he could dispense with this step and proceed to add the sample directly to the acid in his matrass, or test tube, heating it gently. All metals present but gold reacted to form nitrates or chlorides. If nitrates, he then added brine or salt to form chlorides. This produced a reaction in which insoluble silver chloride instantly precipitated as a white, milky cloud which rapidly turned purple-black in the presence of sunlight—the critical test for silver. If that happened, the black precipitate was filtered out, dried, and reduced by blowpipe to a bead of metallic silver. The bead was then gauged by a silver table knife whose blade tip had been sawed or filed into a very acute V-shaped angle (Fig. 5). Cupel buttons of various known weights[23] were borrowed from the local assayer and fitted into the angle to calibrate it at

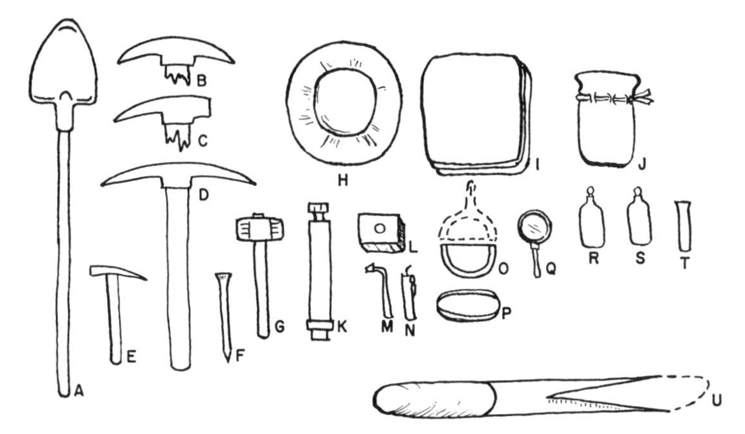

Fig. 5. The prospector's tools. *A*, engineer's shovel; *B*, quartz pick; *C*, poll pick; *D*, common pick; *E*, geologist's hammer; *F*, moil; *G*, four-pound hammer; *H*, prospect pan; *I*, dry-washing sheet; *J*, sample bag; *K*, tube mill; *L*, charcoal block; *M*, blowpipe; *N*, candle; *O*, half-skillet hornspoon; *P*, hornspoon; *Q*, magnifying glass; *R*, acid bottle; *S*, mercury bottle; *T*, test tube (matrass); *U*, knife gauge (enlarged).

the time of manufacture. Rough marks were then scribed onto the blade, with appropriate legends for fast estimation of unknown beads.

Lead and copper content could also be determined qualitatively by a few simple additional tests. Lead chloride would refuse to turn the dramatic black of silver but would remain as a white suspension which could be slowly dissolved in twenty times its volume of water. Copper could be estimated by two methods, the older of which was to dip a bit of clean iron wire into the test glass for a few minutes. If copper was present, it would reveal itself by coating the iron in a clearly distinguishable manner. In the other, later, method crude ammonia poured into the solution would react with the copper present to turn the contents a theatrically bright blue.[24]

Having determined the nature and approximate richness of the mineralization in his outcrop, the prospector now had to consider whether

[23] Though buttons and marks were gauged in "ounces," a "ten-ounce" pellet did not actually weigh ten ounces. Rather, it was the button from a two-ounce sample of ore which went ten ounces to the ton.

[24] Wright, *The Big Bonanza*, 71–72; Wilson, *Silver Stampede*, 31.

or not to take a serious gamble. It was almost always taken for granted that erosion and weathering "milled" and concentrated surface values until these values were considerably higher than those of the unaffected lode itself. The prospector could undeniably sell his claim "as was," but he would not receive a great sum for it, since no one knew what lay beneath. On the other hand, he could clear off the weathered surface to expose the raw rock beneath to the purchaser's gaze and, if it seemed promising, demand and get a much better price. Yet many a good outcrop sat upon a meager vein for which nobody would pay a cent, and there was absolutely no way to determine in advance what lay below. But to be a prospector was almost by definition to be a gambler, and few could resist the temptation to press on. Indeed, a few drill steels and some cans of blasting powder were considered a normal part of the prospector's outfit. Pick-and-shovel work easily broke out the weathered chlorides and rotten quartz at the surface; then two or three drill holes could be put down and shot out to expose the virgin rock. When the tumult died away, the prospector scrambled over to see what nature had provided. All too often it was but a meager stringer of quartz a few fingers wide that would not pay to mine if it were half silver—which it would not be. Now and then, however, the shot would reveal a substantial lead of quartz and sulphurets. At that point the prospector might or might not be in clover.

Normally a prospector only prospected; once he found a promising lode, he desired only to sell it to a mining syndicate, take his money, and go on a prolonged spree at some hog ranch[25] near his base of operations. On occasion, some prospectors knew and employed means of "improving" assay results of their lodes. If, however, the prospector was reasonably honest, he next proceeded to measure the size of his vein and with a pick detach small chips clear across its face to form a representative sample. These chips were to be bagged and carried to what passed for civilization, where they would be submitted to a professional assayer's attentions and disinterested report.

Before departing, however, the American prospector staked his claim,

[25] The hog ranch was a frontier institution not necessarily or even usually concerned with producing pork. Such ranches flourished at the heads of cattle trails and outside military reservations and mining camps. They catered on a commercial basis to the worst in human nature and were redeemed only by their democracy, completely disregarding race, creed, and previous condition of servitude.

a simple common-law procedure. The discoverer of a mineralized ore body gained possession of all the vein, lode, ledge, or lead (these terms being virtually synonymous with each other and with the British term reef), in much the same fashion that a seventeenth-century explorer, discovering part of a river, was entitled to take possession of its entire drainage basin. Until around 1866 the location formula ran about as follows:

> We the undersigned claim three claims of three hundred feet each (and one for discovery), on this silver-bearing quartz lead or lode, extending north and south from this notice, with all its dips, spurs, and angles, variations and sinuosities, together with fifty feet of ground on either side for working the same.[26]

On occasion, the formula also included a parcel of land adjacent to the claim for erection of a mill.

The formula was posted as a notice in the center of the claim, guaranteeing that every partner physically present could claim a parcel 300 by 100 feet for himself and that a similar and additional parcel was awarded as a bonus to the original discoverer in a new mining district. Later the four corners had to be marked with set posts, blazed trees, or stone cairns. In addition, the end lines were staked in their centers—a matter of some moment—and woe betide the scofflaw who wantonly destroyed one. The discovery then had to be described and entered in the ledgers of the local registrar of claims within a specified time, usually thirty days. Location work equal to one hundred dollars' investment[27] had to be performed within ninety days to validate the claim for the first year, and thereafter the same amount of assessment work had to be performed annually. Otherwise, the claim would be regarded as abandoned and would be subject to location by the first person to come upon it subsequently.

It was customary for the prospector to christen his location; in fact, local registrars often indexed mineral claims alphabetically according to the names given to them. Prospectors were prone neither to doctrinaire egalitarianism nor even to modesty in this matter, as shown by the profusion of imperial, monarchic, and plutocratic names that prevailed

[26] Clemens, *Roughing It*, 211.

[27] A general rule of thumb was that the location shaft should be sunk at least ten feet on the ore body.

throughout the mining West, such as the Crown King of Arizona, Midas of Nevada, and Sam Clemens' own Monarch of the Mountains. An attractive name was as useful in promotion as was a good ore showing—or perhaps even more so.

Humor could be expressed or old scores evened by such legally sanctioned means. A case in point is that of Almeda Reed Shoemaker (1861–1938), who as a little girl in the Black Hills had a promising lode claim, the Little Allie, recorded in her honor by a friendly neighbor. The neighbor's wife reportedly grew resentful and said to him, "You've named claims for your favorite politicians, for your old girl friends, and even one for your pack mule. Now, I've looked it up, and I'm told you can amend the location to change the name, and I demand that you do just that and name this one for me!" Her husband gave in and recorded the amended location as the Holy Terror, which name it still bears.

CHAPTER II

Assaying and the Art of Salting

Having marked his claim by one of the methods described in Chapter I, the prospector now made his way back to the settlement. There he registered his claim, paying a small fee for the service, and then hunted up the nearest custom assay office. The "professional" (now "registered") assayer was a man of wide parts, a practitioner, after the physician, of the world's third-oldest profession. (Scientifically speaking, he was even ahead of the doctor; frontier medicine was still hampered by quackery, but the assayer maintained a proud tradition which had long before cast off its somewhat disreputable association with alchemy.)

Assaying originated with the ancient Egyptians, was enlarged by the Arab philosophers, and was brought to perfection by the methodical German experimenters of the fifteenth and sixteenth centuries, whose works were disseminated by the great textbooks of Bauer and Lazarus Ercker.[1] Barring a handful of technicians of the United States Mint, there seem to have been virtually no assayers in the nation in 1849. The lack was quickly remedied, however, by alumni of the Royal School of Mines in England and of the Freiburg School of Mines in Saxony. For a generation thereafter the art was probably transmitted by apprenticeship training. Indeed, an assayer can hardly manage without an apprentice or two, since much of his activity involves time-consuming physical labor of which any intelligent youth is fully capable.

Custom Assaying

The work of the custom assayer consisted in receiving samples of mineralized rock delivered to his door by prospectors and mine investigators. He would subject these samples to a quantitative analysis, either by

[1] Georgius Agricola [Georg Bauer], *De re metallica* (trans. and ed. by Herbert Clark Hoover and Lou Henry Hoover), Book VII, 219–65. See also Lazarus Ercker, *Treatise on Ores and Assaying* (trans. and ed. by Anneliese Grünhaldt Sisco and Cyril Stanley Smith), 37–38, 131–33.

"wet" chemistry or by the fire assay, which was a highly refined sort of smelting. The assayer then reported what metallic elements were in the sample and in what quantity, as requested by his customer. He received an established fee for each determination, presumably confining himself to a mere finding of fact on the evidence presented. In actuality, tens of thousands of dollars might change hands on the basis of his conclusions, wherefore it was assumed that the prudent assayer combined in himself the detective ability of a Pinkerton, the taciturnity of a Sphinx, and the worldly wisdom of a midwife. He was, in short, an educated man and a pillar of the community, drinking from the same jug with mine superintendents and bankers. He might not be an alderman, but he was almost invariably a member of the board of education.

The assay office was usually a one-story frame or adobe structure, marked by the inevitable roofed-over front porch, piled high with bagged ore samples awaiting assay. Within, it was divided into three rooms, of which the front room was the office, equipped with desk, cupboards, shelves of reference books and mineral specimens, candle boxes and tin cans full of other specimens, prospecting equipment, and, nailed to the wall, such works of art as men of technical persuasion invariably receive from the jobbers who solicit their orders. The assayer was seldom in this office, nor did he encourage loitering about the premises. A bell connected to the front door announced the arrival of customers, who knew that their wait for the assayer to appear might be a lengthy one.

The room behind the office was the laboratory, prominently featuring the two-decked assay furnace of iron-banded fire clay, work tables, such wet assaying glassware as might be needed, crates of crucibles, shelves of cupels, and other implements of trade. Carefully partitioned off in one corner was a tidy balance room, or cubby, as dustproof as it could be made. This room contained only chair, table, glass-cased balances, and superior works of art reserved for the assayer's moments of private contemplation. The back room was a dismal place, unkempt and ill-lit, containing the ironmongery used for crushing and screening ore samples, molding equipment for forming cupels, and supplies. Over everything lay a heavy film of rock dust. The most constant and pervasive sound was the roaring of the assay furnace, to which the apprentice's labors added a cacophonous obbligato.

The greatest part of the work lay in running fire assays for the determination of the gold-silver content of ore. After the assayer received the sample, he turned it over to the apprentice, who was charged with pulverizing it and reducing its bulk to manageable proportions. To this end he first employed a hand-turned chipmunk, or miniature jaw crusher, to reduce the sample to pieces the size of grains of corn. If the sample was quite bulky, it was then coned and quartered: it was thoroughly stirred up and heaped into a cone, and the opposite quarters of the cone were removed and discarded until the pound or so of particles remaining could be considered truly representative of the sample average. These particles were then poured into a "fatter" mortar, and its handle was energetically spun by the apprentice until the particles were reduced to sand and dust size. Using a one-hundred-mesh round wire sieve, he would shake down the dusts and then dump the oversize, a tablespoon at a time, onto the bucking board, a flat, cast-iron plate about two feet square, with both sides turned up an inch or so and both ends open. The oversize was worked down with a muller, a cast-iron, curved object weighing about twenty pounds and fitted with an ordinary ax handle. The muller was pushed and dragged back and forth with a rocking, dragging motion until the oversize was reduced to a size that would pass through the screen. Having delivered the prepared sample to the assayer, the apprentice stoked the furnace, carried out ashes, and began work on the contents of another sample bag.

Using moderately sensitive balances, the assayer carefully weighed out a fixed and arbitrary portion (then a trifle over 2 ounces; today 29.116 grams) of the sample and poured it into his crucible, a tapering cup of fired and glazed china with a capacity of about 1 liquid pint. To the sample he added litharge (lead oxide), borax, argol, wheat flour, and (when indicated) an inquart needle of silver of precise purity and weight. After stirring all these ingredients together, he placed the crucible into the lower, or reducing deck of the furnace, where it was subjected to a temperature of about 1,500° C and an atmosphere of carbon monoxide calculated to draw out the oxygen from the litharge in the sample. Tightly shutting the D-shaped door, he thereby encouraged the litharge to be reduced by the flour and the furnace gases to droplets of metallic lead, which would descend through the melt, collecting, fluxing, and smelting whatever precious metals might be present. The process took about half an hour. Then the assayer reached into the furnace, seized the crucible with the fork of an iron crutch[2] and poured the glowing melt into a heavy cast-iron mold with conical receiving cups. The melt poured in a golden-white run. About two-thirds of the way through, the molten lead, looking like a small, silvery minnow, would slip swiftly over the crucible rim.

When the melt had cooled into a conical biscuit, the assayer knocked it out of the mold and began smiting it lustily with a hammer. The process broke away the glassy slag, leaving the conical tip of lead. He pounded this little cone into a rough cube approximately 0.5 inch wide on each face. While he was doing this (for he usually ran as many as a dozen assays at once), he was preheating the cupels in the upper muffle, or oxidizing deck, of the furnace. The cupels were molded of moist bone ash (in Bauer's time, of the ash of deer antlers) in the back room; they were neither fired nor glazed, nor could they be used more than once. The cupel, resembling an old-fashioned salt cellar in size and shape, was about 1.5 inches in diameter and 1 inch high, with a hemispherical depression, or cup, in the top (Fig. 6). The depression was large enough to accommodate the lead button when it melted, while exposing the maximum surface. It was necessary to preheat the cupel to drive off the molding moisture; otherwise, when the moisture came in contact with the molten metal, it would spatter deplorably.

Using long forceps, the assayer deposited each lead cube in its glowing cupel. This part of the process was carried on at a lower heat (900° C) than that of the earlier process. The cupels were shielded from direct flame by a semicircular clay arch, the muffle; the D door was left partly open to ensure a supply of oxygen to the rapidly melting button. The oxygen combined with the lead, returning it to litharge. This process was critical. Some unknown experimenter had discovered a unique virtue of bone ash: it was impervious to molten metal but would absorb litharge like a sponge. During this process of oxidization and absorption the assayer kept a close eye on it. Too great heat would volatilize some of the gold and silver, while too little would allow the button to "freeze," or solidify; thus the heat had to be kept above the melting point of litharge but well below that of any combination of gold and

[2] Hence the shape of the crucible, unchanged since the times of the Pharaohs, which permitted it to be safely grasped and tilted with the crutch or, at an earlier date, with two metal or wooden sticks.

Fig. 6. Renaissance equipment for washing and molding bone ash for assayer's cupels. *A* and *C*, molds for cupels; *B* and *D*, cupels; *E*, stack of cupels; *F*, ball of washed ashes; *G*, man washing ashes; *H*, man pounding cupels. From Lazarus Ercker, *Treatise on Ores and Assaying* (1951 translation).

silver. Since this observation was hard on the face and eyes, the assayer watched the process through a slit in a board equipped with a handle. This board, or tablet (later replaced by smoked glasses and later still by a welder's mask), was as much the trademark of the assayer as were the mortar and pestle those of the druggist (Fig. 7).

Fig. 7. A Renaissance assayer watching cupel through a tablet. From Georgius Agricola [Georg Bauer], *De re metallica* (1555).

Cupellation took another half hour, during which time the glowing pool of litharge subsided slowly but steadily until the watchful assayer determined that it had vanished. Then with a miniature iron hoe he nursed the cupel onto a flat spatula and removed it from the furnace. Amid its fading glow were visible the yellow "feathers" of litharge crystals and a tiny red sphere; the latter was the doré (gold-silver) bead, which would presently "blick," or give up its latent heat in a rather spectacular flash of light as it solidified. This was in many respects the moment of truth. The inquart silver needle produced a tiny bead of well-known size. If the cupel bead was no larger, the ore sample was valueless. Should it be any larger (assuming that all copper had been earlier fluxed out), the sample had an appreciable amount of precious metal.

The doré bead was cooled, brushed clean, flattened, and weighed with scrupulous care on the assayer's most sensitive balance. It was next put into nitric acid, in which it was heated gently. Now the inquarting bore fruit, for the correct proportion of silver to gold was (and is) four to one for the most effective acid parting, and must be adjusted to that ratio, if necessary, by the addition of the silver inquart needle. Too high a proportion of gold would prevent the acid from doing its work, while too much silver would make the gold particles fine and difficult to collect. If the parting went well, the silver dissolved to leave the gold as a fine, sootlike, amorphous body. It was filtered, washed, dried, and annealed in a porcelain vessel, upon which it reassumed its characteristic appearance. It was then weighed and the loss of weight (corrected by subtraction of the added inquart) was the silver content of the sample. Later, in his office, the assayer transferred his results to an official-looking certificate. In early days the determinations were expressed in dollar values of gold and silver per ton of ore and, after about 1873, in troy ounces of precious metals per ton.[3]

It should be noted that the base metals were ignored in this sort of assay and that no attention was paid to any values collected by the slag. Despite this weakness in the method, the fire assay for long reigned supreme; milling methods were unable to extract values which the fire assay missed, and often enough half the gold-silver content of a refractory ore went down the flume to the tailing heap outside the mills.

Mine Assaying

Custom assaying was but part of the story of this craft. Mines also often required the services of resident assayers. It was supposed by the laity that lodes of precious-metal ores were quite distinct from the country rock in which they are found and that ore could be readily sorted from the worthless waste produced in drifting. In a great many cases this theory was correct, but in a great many others the ore values were not distinctly visible but thinned out gradually from the lode into and through the country rock, often as the result of water action. Wet mines were notorious in this respect, but even very dry mines could be affected. Since not even the expert eye could differentiate ore that was worth milling from ore that was too thin to mill profitably, progress was determined and checked by a continuous series of assays run at the headworks. The assayer gave the results to the shifter, or shift boss, who then marked the spots for the next round of drift shots accordingly, edging the drift more into the pay ore and away from the thinner rock (Fig. 8).

[3] I am indebted to Walter Statler, R.A., of Humboldt, Arizona, who kindly demonstrated a classical fire assay for my edification.

Fig. 8. The shift boss. Reproduced by permission of Buck O'Donnell and Shaft and Development Machines, Inc.

Salting

With location patent in one hand and assay certificate in the other (and, usually, a consuming thirst in between) the prospector now went in search of a buyer at the central oases where such persons congregated. When the two met, the proceedings were of a nature to shame a Levantine rug peddler. The prospector inflated the value of his claim with all the art and forensic cunning in his power; the buyer automatically deprecated the prospector's words and his assay. That the assay was professionally certified was a fact to be gently waved aside. The buyer who did not personally visit the lode to get his own samples, plus an independent assay report, would be regarded by the community as a cretin; *every* discovery was considered salted until proved otherwise beyond reasonable doubt.

The very mention of the word salting in mining country will to this day provoke cries of indignation from prospectors or descendants of prospectors, who like the modest lady to her swain will state firmly that they had nothing of that sort in mind, and besides they do not know what you are talking about. Accordingly it is perhaps best to let the experts themselves speak on the subject. These men were for the most part experienced consulting mining engineers of vast probity and knowledge, if not morose cynicism. For example, Arthur Lakes, of the Colorado School of Mines, said flatly, "It often happens that a miner, who in every other relation of life is honest as the day, draws a line, when it comes to the selling of a mine, which he considers 'fair game.' "[4] Thomas A. Rickard, the dean of American precious-metals mining from 1890 to 1929, wrote with equal pessimism that "an engineer when about to examine a mine should write *caveat emptor* on the first page of his notebook."[5] Later he quoted Bret Harte's couplet: "The ways of a man with a maid be strange, yet simple and tame/To the ways of a man with a mine when buying or selling the same."[6]

The best way for the prospective buyer of a mining property to protect himself against mischief was to deputize a consulting engineer to visit the place and make a thorough investigation and report. Such an engineer had much to look out for. Lakes reports that the entire mining community might enter a general conspiracy against him, on the consensus that an encouraging report would be to the economic advantage of all in the community whereas a negative report would be good for nothing. Therefore, the engineers were treated like visiting royalty; Lakes hints that the first move in the game might be to salt the expert himself by befuddlement as well as bribes. This communal feeling went

[4] Lakes, *Prospecting*, 271.

[5] Thomas A. Rickard, *A History of American Mining*, 392–93.

[6] Thomas A. Rickard, "Salting: The Nefarious 'Art' and its Age-old Results as Practiced on the Mining Fraternity," *Engineering and Mining Journal*, Vol. CXLII (1941), 42 (hereafter cited as "Salting").

so far that the miners themselves might salt the mine quite independently of the seller's intentions, so as to secure continued employment (Fig. 9).[7]

Fig. 9. "The bully honest miner." From William Wright [Dan deQuille], *The Big Bonanza* (1964).

The most famous example of successful salting was surprisingly untypical of most cases. This was the Great Diamond Hoax, in which two confidence men, Philip Arnold and John B. Slack, about 1870 salted a Colorado mesa with a large number of flawed, industrial-quality, or otherwise inexpensive gem stones of all sorts. This geological solecism alone should have been their immediate undoing, for different varieties of precious stones rarely occur in the same surroundings. Arnold and Slack had the tact, or the good fortune, however, to win ascendancy over the mind of Henry Janin, then considered one of the foremost mining experts in the United States, even though he had no experience to speak of concerning gem stones. Apparently Janin so lusted for professional recognition that he swallowed the story whole after being conducted in theatrical secrecy to the site. Using his report, Arnold and Slack coaxed large sums of money from the Rothschild interests and from William C. Ralston, of the Bank of California. Ultimately, Clarence W. King, of the United States Geological Survey, grew suspicious, found the salted mesa, and in 1872 published a withering exposé of the fraud. Although King's report emphasized the fact that rubies and diamonds are not found together in nature and that none of the stones was recovered from unimpeachably natural surroundings, his suspicions are said to have been first confirmed by the finding of one diamond with a jeweler's facet already polished upon it.[8]

The Great Diamond Hoax was unique. "Normal" salting almost always concerned itself with the precious metals (though high-grade copper ore was also a great favorite) and as a rule ran to two general patterns. The first and crudest of these patterns was to salt for the benefit of persons without much experience, native intelligence, or perceptible desire to secure the opinion of a qualified expert. The most primitive form of this type of salting was the practice of searching over a comparatively worthless ore ledge in order to secure one specimen of higher grade, which would then be submitted for assay and its values declared typical of the whole.[9] Those who bought a mining property on the

[7] Lakes, *Prospecting*, 265.

[8] Rickard, *A History of American Mining*, 380–96, and especially 395. See also Asbury Harpending, *The Great Diamond Hoax and Other Stirring Events in the Life of Asbury Harpending* (ed. by James H. Wilkins), *passim* (hereafter cited as *The Great Diamond Hoax*); Bruce A. Woodard, *Diamonds in the Salt, passim.*

[9] Clemens, *Roughing It*, 216.

strength of the seller's estimation after this fashion were regarded as unworthy of pity.

A slight improvement in technique was found in the attempt to salt a nondescript Comstock mine, the North Ophir, with melted silver half dollars. These lumps were merely blackened to resemble native metal, then mixed with the dull-yellow syenite country rock of the lode at the bottom of a short shaft. But the Washoe speculators were in a receptive mood, and the North Ophir experienced a brief flurry of interest until "-ted States of" was found on one of the lumps of native silver. Like the facet upon Janin's diamond, this evidence of sophistication instantly deflated the boom.[10]

Farther up the scale was the practice of scattering ore of good tenor obtained from a going mine about the outcropping or lode face of the property for sale. A lazy man would gather these loose specimens for assay, then purchase the property at the true valuation—of another mine. The classic case of this sort of fraud resulted, however, in the biter being bitten. In 1875 it was discovered that the Leadville limestone of central Colorado was very profitably mineralized with silver and lead. This was a thing heretofore considered impossible, but the strike excited burning interest in Colorado limestone formations of all sorts. One Chicken Bill Lovell got some of the Leadville ore and dumped it about an unattractive limestone outcrop, which he named the Chrysolite, and he proceeded to sell at a fancy price to H. A. W. ("Haw") Tabor—a man who exactly fitted the specifications already given for a desirable mark. To Chicken Bill's dismay, Tabor innocently proceeded to develop a very profitable silver-lead operation upon the salted outcrop.[11] Normally, though, this trick was exposed by the fact that the imported ore would seldom if ever have characteristics identical with those of the salted ledge; besides, a careful appraiser never relied solely or even partly upon grab samples, except as a check upon his other specimens.

A more imaginative approach to the problem of inflating real but thin values was worked out by early Washoe, Nevada, placer men about 1858. They were working Six-Mile Canyon on the slopes of Mount

[10] *Ibid.*, 311–12. Those who are inclined to dismiss this tale as one of Mr. Clemens' exaggerations will find it abundantly confirmed by Rickard in "Salting," *Engineering and Mining Journal*, Vol. CXLII (1941), 42.

[11] Rickard, *A History of American Mining*, 130.

Davidson but recovering only bogus gold (so called because the dust was extremely pale, the color almost of silver—as indeed it was, although no one yet realized the fact). The itinerant clean-up buyers of the district declined to purchase this dust until the placer men ingeniously adulterated it with California gold dust of healthier hue, thus getting a gold price for their gold-silver accumulations.[12]

Quite similar was the custom of selling a worked-out or thin gold placer by the simple device of ensuring that enough color went in with the dirt at the head of the sluice to guarantee an attractive cleanup of the riffle bars at the end of the demonstration run. A prospective buyer might try hard to keep an eye on all the proceedings, but he could not be everywhere at once—in the pit, at the head of the sluice, and along the transportation path. Provided only that the characteristics of the salt gold matched those of the claim in color, shape, and size, the swindlers might well get away with it. But it would never do to have two different sorts of gold dust emerge from the sluice, as might be the case when gold-coin filings of spicular shape and reddish hue were used to salt a placer which held rusty dust or lenticular scales. Each of the partners in crime having some dust of the right sort to distribute where it would do the most good—this one in the ashes of his pipe bowl, that one beneath his fingernails, and a third man in some mud carelessly smeared on his shovel[13]—the odds were strong that an unsuspecting buyer might be taken in, particularly if the placer had reputedly produced well in times past.

Inexperience and haste on the part of the buyer were always the allies of a salter, who was not invariably the owner of the property. As has been noted, the miners themselves sometimes salted ore they were mucking after a survey blast of a lode face, using "ruby" silver or liquid silver nitrate to improve the assay of silver ore—or gold dust scattered across the muck pile from a small hole in the pants pocket, if a gold mine was in issue. Or the miner might secrete some dust in the band of his hat. Ore samples traditionally were collected in the hat before being placed in a sample bag, and the simple act of inverting the headpiece for this end would shake down enough dust from inside the band to show a greatly improved return.[14]

[12] Eliot Lord, *Comstock Mining and Miners*, 36–37.

[13] Rickard, "Salting," *Engineering and Mining Journal*, Vol. CXLII (1941), 50.

[14] Thomas A. Rickard, *Journeys of Observation*, 209.

In all these examples there is one point—and often two points—in common. In every instance the purchaser or engineer neglected to be certain that the placer dirt or ore was exactly as nature had deposited it. Furthermore, the frequent indolent habit of permitting people who had an interest in a favorable report to handle the material was simply asking for it. The wary mining engineer quickly learned to take due precautions in both these matters.[15]

The second and more sophisticated pattern of salting included certain types of placer and mining operations which were potentially dangerous even to experts because the salting had been especially intelligently planned. The salter of wide parts did not try to overdo the richness of his property, and he used methods far superior and more insidious than the crude sleights so far mentioned. To begin with, such a salter preferred to start with a real gold mine or placer of once-profitable tenor. This practice ensured that the geology would by definition be right. He preferred to confine himself to gold producers, since they were intrinsically easier to salt than were silver lodes, for two reasons: a little gold was far easier to manipulate than the much larger quantity of silver necessary to give the same dollar per ton upgrading, and gold was often found in simple compounds or even the native state, whereas silver tended to form complex ores which were very difficult to imitate convincingly. In other words, a grain of gold in the right place might salt a mine with an ease which several pounds of silver nitrate could scarcely rival. This is not to say that silver mines were never subject to salting, but only that an engineer had to be particularly alert when investigating a gold property.

Salt gold usually came from three sources: natural gold dust, coin filings, and soluble gold chlorides (or, at a later date, gold cyanides). Gold chloride was a great favorite among evildoers. In the 1870's and 1880's it was included in patent medicines widely advertised for the relief of kidney complaints stemming from the immoderate use of whisky. For obvious reasons, such medication was readily obtained in the West. One unverifiable story tells of a thrifty salter who actually had such a kidney complaint. He took his medicine, then transferred his own assay values to the surface of his lode as the spirit moved.

Liquid salts could be poured into crevices, drill holes, and the like

[15] Rickard, "Salting," *Engineering and Mining Journal*, Vol. CXLII (1941), 50.

where solids could not be convincingly introduced. On the other hand, soluble gold salts are never found in nature. Suppose that an assayer had determined by a fire assay that a sample contained 20 ounces of gold to the ton. Suppose, further, that he, like the Irish Member of Parliament, smelled a rat, saw it floating in the air, and proceeded to nip it in the bud. To wit: suppose he washed another sample from the same batch, reassayed it, and found that it went only 0.5 ounce to the ton. The inferences would be plain enough, and it would be taken for granted that the presence of gold chlorides was prima-facie evidence of attempted stratagems and spoils.[16]

Coin filings were very often used for unsavory purposes, particularly in the Black Hills mineral districts, where professional assayers seemed to be scarce and surveys were made chiefly by the crude method of sending a wagonload of mine-run ore to a mill for evaluation.[17] When slipped quietly into mill mercury, coin filings would fatten up the yield as well as anything, but when used to salt rock samples, their spicular shape and reddish color seen beneath a microscope tended to betray their origins. If subsequent careful analysis revealed the copper content of the gold particles to be identical in proportion to Mint statutory percentages, there was little need to investigate further.[18] By elimination, therefore, natural gold dust was the favorite instrument of vice, and it was frequently used to salt both placers and lodes.

The simplest way of salting a placer was to sow gold colors along the edge of the stream, trusting that hydraulic stirring of the sands would soon carry the salt down to bedrock to form a modest pay streak which the digging of a prospect pit would reveal. This method was especially promising if bedrock was comparatively near the surface, but was often betrayed by the unnatural uniformity of gold distribution; in nature, the pay streak tends to be spotty. An improvement was to dig small pits down to bedrock in places likely to be sampled, enrich the pay streak, and refill, being very careful to replace the spoil with a graded mixture of sand and pebbles which would leave no telltale profile when the sample pit was excavated.[19]

[16] Lakes, *Prospecting*, 267.

[17] Hughes, *Pioneer Years in the Black Hills*, 289–90.

[18] Rickard, "Salting," *Engineering and Mining Journal*, Vol. CXLII (1941), 42.

[19] Hubert Burgess, "Anecdotes of the Mines," *Century Magazine*, Vol. XLII (1891), 269.

Although gold dust could be sown, dribbled, poked, injected, and plastered on surfaces with clay, its nearly universal instrument of application was the shotgun. The cheap, reliable scatter-gun was found all over the West. It was the favorite weapon of both defense and offense; to possess a shotgun was far less remarkable than to be without one. It had in addition the unique virtue of being easily loaded with gold dust and the capability of blowing that gold dust into a placer or lode face in a manner remarkably deceptive even to the most knowledgeable of consultant engineers. From that day to this, mining textbooks have had little good to say about this "gun that won the West"—or at least salted a good deal of it.[20]

Let Hubert Burgess (doubtless a disinterested party) tell of a group of American placer men who in 1851 were working an extensive but thin placer near Columbia, California. A syndicate of Chinese expressed some interest in acquiring the claim, and the Americans naturally decided to salt it in order to get a better price. Their problem lay in deciding where to put their salt gold, for the very size of the claim prohibited general salting such as has been discussed. At length one of the Americans decided to observe the Chinese at their survey panning, then salt between pans, right under their noses. To this end he found and killed a large snake and then gave it to a confederate with orders to remain on the bluff overlooking the claim during the survey. At a given signal the snake was to be tossed over the edge of the bluff, so as to land virtually on the surveyors' heads. The rest would follow with ease, since the chief schemer would be sauntering along behind the panners, shotgun in hand, and with an appropriate load in the gun. Burgess's tale concluded:

> Next morning, four Chinamen came prepared for work. They tried a few places, but of course did not get the "color." The Americans kept at a distance so there could be no complaint.
>
> "Well, John," said the schemer, "where you try next, over in that corner?"
>
> The Chinamen were suspicious in a moment. They were familiar with

[20] Since proverbial facts are often found to be not facts at all, a selection of authorities is offered to substantiate the wide and frequent use of shotguns in salting: Lakes, *Prospecting*, 266; Rickard, "Salting," *Engineering and Mining Journal*, Vol. CXLII (1941), 43; Hugh E. McKinstry, *Mining Geology*, 67; Robert S. Lewis, *Elements of Mining* (3d rev. ed. by George B. Clark), 358.

salted claims and were well on their guard. "No likee dis corn'. Tlie him nudder corn'," pointing to the opposite one.

Jim, with his hands in his pockets, was above on the bank, many feet away, watching; when he saw them point in that direction, his partner gave a nod and he pitched the snake on the ground near the place. The leader exclaimed, "Hold on boys!" and fired before they could tell which way to look. Going up to the snake, he pushed the gun under it and carried it away hanging over the barrel. Jim walked off and Bill sat on a wheelbarrow on the opposite side from where they were at work. The Chinamen had no suspicion. They carried away several pans of dirt to wash in a stream near by, and when they returned Bill felt pretty sure they had struck some of the "salt," but the Chinamen said nothing except, "Claim no good. Melikin man talkee too muchee."

The Americans, knowing the game, refused to take less than the specified price, which the Chinamen finally paid and in two days the sellers were off to new diggings.[21]

Salting a mineral lode by shotgun was not greatly different, except that one might do without the snake. One mine owner, however, tried to dispense with the shotgun as well. The story (unverified) has it that he had his property properly examined, and since it had little value, a negative report was submitted. The owner took thought and presently invited the engineer to return for another look, saying that he might have "overlooked something." The engineer good-naturedly returned but soon noticed that pats, smears, and gobbets of clay had mysteriously appeared throughout the workings during his absence. Having a pretty fair idea of what the clay contained, the engineer sweetly asked the proprietor whether it had lately been raining *very* hard in his mine. The reply has not been recorded.

An answer to shotgun salting of lodes was quickly evolved. Since nature usually deposits mineral from the inside out and the salter applies it from the outside in, it followed that shotgun salting at best merely blasted a thin veneer of gold specks into the face of the lode. The routine countermeasure was to demand that the face be drilled and shot out, so as to expose completely virgin rock for sampling. It also followed that the survey engineer hung about during the drilling, with a view to preventing the miners from dribbling chlorides or nitrates down their

[21] Burgess, "Anecdotes of the Mines," *Century Magazine*, Vol. XLII (1891), 269–70.

holes, under the guise, say, of casting in water to remove the rock cuttings as a mud clinging to the drill steels. After the blast was fired, the engineer re-entered the drift alone to take his samples from the new face.

Despite this apparently foolproof procedure, it often became manifest that the new faces were salted. Worse yet, the salters were doing it *in absentia* and with gold dust. This seemed impossible, but at length the trick was exposed. The holes were honestly drilled, but the ends of the headsticks of the dynamite charges were loaded with salt dust. The blast distributed the gold over the new face, driving it into the vein rock in a thoroughly deceptive fashion. After the trick was discovered, survey engineers furnished their own dynamite.[22]

Less ingenious salters realized early in the game that, even if they could not "improve" their in-place ore, there was little to prevent them from tampering with ore samples in transit to the assayer's office, or even at the office itself. Half the lore of salting concerns the success or failure of the engineer to guard the pristine purity of his sample bags.

T. A. Rickard related seventeen major cases, of which eight dealt with just this aspect of the problem. He told of the successful salting of Professor James Douglas, developer of the famed Copper Queen Mine at Bisbee, Arizona. Douglas became interested in some Gila River copper property owned by a freighter named Tweed. After a favorable report on the claim by his associate, Benjamin Williams, Douglas bought the property on the account of Phelps Dodge, only to discover after some expensive development that the ore was almost valueless. Inquiry revealed that Williams had kept his samples in a carpetbag whose leather bottom was in disrepair and had been roughly mended with twine. While Williams slept, Tweed entered the locked bag by cutting the twine. He judiciously enriched the samples with powdered malachite, then sewed shut the rip with the same kind of twine. To make matters worse, Douglas himself had taken further samples, using Williams' carpetbag, and had been served the same way by the ingenious freighter.[23]

With gold or silver properties, all that was necessary was to gain momentary access to the sample bags and then with a syringe loaded

[22] Lewis, *Elements of Mining*, 358; Rickard, "Salting," *Engineering and Mining Journal*, Vol. CXLII (1941), 54.

[23] Rickard, "Salting," *Engineering and Mining Journal*, Vol. CXLII (1941), 52.

with gold chloride, silver nitrate, or gold dust do one's worst. For this reason engineers tried to keep their sample bags in full view from lode face to assay office. Even so, they might be taken off guard by clever misdirection. A favorite form of this trick was the "drunken stranger" gambit, in which a hilariously tanglefooted inebriate would hiccup, lurch about, deliver himself of crapulous witticisms, and in general get the engineer laughing so hard that he would scarcely notice that the drunk had fallen and smashed his whisky bottle over the sample bags or wagonload of test ore. Needless to say, the drunken stranger would not be at all drunken, nor would his bottle contain whisky.[24] A good countermeasure when en route tampering was anticipated was to include "blind," or worthless, ore samples, bagged and tagged as though they were mine-run specimens. If subsequent assay showed them to have substantial values, something was obviously amiss.

Even that stronghold of scientific probity, the assay office, was not always above suspicion. A garrulous visitor might wander into the laboratory room at the psychological moment, smoking a pipe or sporting remarkably dirty fingernails. At a moment's distraction of the busy assayer, the pipe ashes or fingernail contents would disappear without detection into the crucible. By that method a few tiny specks of gold could go a very long way. More determined souls might even burglarize the office to add salt mineral to the assayer's fluxes. Engineers suspecting such huggermugger were known to chase the assayer out of his office, install their own locks, and conduct the assays themselves.[25]

Now and then there were some remarkably ingenious variations, one of which could be said to consist of salting without salt. Lakes told of a silver mine of good tenor which was routinely offered for sale. It was quite carefully surveyed, found to be without flaw or blemish, and vended with no more than the usual amount of haggling. The new owners arranged for production to start and sat back to count their profits. They did not do so for long, for after the first face blast, the miners boiled up to the surface to report in astonishment that they had "holed through" into an unsuspected continuation of the drift. A brief inspection showed the infuriated purchasers that the vendor had stopped work at the face while his mine was still going strong. He had then run

[24] Hughes, *Pioneer Years in the Black Hills*, 298–99.
[25] Thomas A. Rickard, *Retrospect: An Autobiography*, 61–72.

in a cleverly concealed bypass winze and taken out the rest of the good ore, leaving the original face in place as a thin but natural screen. The meager contents of this screen were all that the purchasers realized on their investment.[26]

The Growth of the Mining Districts

Given a reasonable degree of probity and a fair number of placer strikes or ore showings in a given district, a boom would soon be under way. Other prospectors would come in, and often would secure temporary work in going mines or mills, both to build up grubstakes and to secure a free education in the profitable geology of the district. Having secured enough of both, they would draw their time and depart for the hills. Prospecting in a new mineral district proceeds naturally from the low country to the back-country highlands and evolves from placering to hard-rock mining. Lowland placer diggings are the easiest for the first-comer to discover and work, since so much of the preliminary labor has already been accomplished naturally. Nonetheless, the temptation to press upstream is great. When prospectors find a big nugget of gold, they can barely stay long enough to eat lunch, since they conclude that the lode of origin is both near to hand and richer than the proverbial yard up a bull's throat. Neither assumption is necessarily correct. For one thing, nuggets are said to "grow" in a stream bed by a process of melding and accretion. For another, a thin lode can erode off over the years, to produce by natural concentration a very attractive high-grade placer at its foot. Similarly, an outcropping stockwork vein system, such as that at Rich Hill, shot through with threadlets of gold, might assay hundreds of dollars to the ton, yet not pay to work because of the vast quantities of gangue present, but still provide high values in native metal upon its outcrop. Despite such strictures, a placer district inevitably tends to expand upstream, and therefore it will not be long until someone is led to the lodes from which the placer colors originated.

Thus, in the beginning the metals were there: gold in the Sierras, the Rockies, and the Black Hills, copper in Montana and Arizona, silver in the Great Basin, waiting for the heirs of Europe to come and wash or dig them out. The Indians' ascendancy was over the land for fourteen thousand years, but the Indian did not prospect for metal. Stumbling

[26] Lakes, *Prospecting*, 270–71.

Mining zones of the West

over it, he did not recognize it for what it was, or speculate about what it might be good for beyond ornamentation. The Amerind made a number of false starts toward a Bronze Age but somehow missed the vital connection that was made repeatedly in the Mediterranean world and in China. What was the absent element?

It has been hypothesized that, all else being reasonably equal, the Amerind had a quite adequate supply of flint and other such silicious rock, eminently suitable for Stone Age toolmaking, and therefore rested content with what he had. Native copper or gold nuggets, available for the taking, could be hammered into pleasing shapes or even utilitarian objects, but these metals would be inferior to silica and so were put aside. In Mesopotamia and China stone is hard to come by, being buried under a huge overburden of loess or alluvial silt. Egypt has granite, limestone, and sandstone in plenty, but none of these will do for sharp-edged tools. Hence the prehistoric inhabitants of these lands would seize upon copper as better than nothing, desire ardently to improve upon it, and be presently launched upon mining, metallurgy, and smithing. What they learned would then be disseminated to the ends of the known world, in proof of the historical truism that occasionally a little deprivation and poverty is no bad thing.

CHAPTER III

The Spanish Tradition

UNTIL THE LATE 1880's the western mining frontier depended so heavily upon Spanish milling methods that it is almost correct to say that millmen used little else but Spanish techniques hooked up to steam engines. Western placering methods may have originated in the state of Georgia and mining methods in Cornwall, but milling and many related activities, such as chloriding and the working of desert lodes on a shoestring were borrowed from Mexico and its three centuries of experience in adapting Renaissance developments to a land rich in minerals but lacking much else.

The high, arid plateau of the Mexican silver belt (a geological continuation of the Great Basin silver lodes yet to be prospected) was relatively devoid of fuel, timber, and waterhead, all essential to effective preindustrial mining. The silver ores outcropping on this plateau proved after a short time to be of a sort which were singularly difficult to reduce to metal. Mexico also lacked the coal, iron, lead, and mercury which were indispensable to contemporary metallurgy. The Spanish crown was characterized by procrastination and a degree of economic misrule. The mine operators of Mexico disposed only of what appeared the most backward subjects to educate in a recondite and intellectually demanding profession. Of this depressing catalogue of difficulties, some were overcome by brute force, some were resolved by extraordinary ingenuity, and some were merely evaded—all with such speed that, a mere half century after the Conquest, Mexican silver ore was being broken, hoisted, reduced, and minted. Many of the techniques so developed would spread north to the American mining frontier at mid-nineteenth century.

Origins

The Spaniard brought to Mexico rather more than the cross and the sword. He brought as well a mining tradition founded on the Carthagenian and Roman exploitation of the Iberian gold, copper, silver, mercury, and iron deposits. The classical technology was Levantine in origin, and, although Iberian skills suffered regressive eclipse during the Visigothic and Moorish invasions, they were never totally extinguished. A striking renaissance in mining technique then began with the accession of the Emperor Charles V to the throne of Spain in 1519. Imperialist interests opened Spain to the influence of the highly skilled Saxon mining technicians; it is known that such Saxons were soon living and teaching in metropolitan Spain.[1] In the year following Charles's abdication the first and greatest of all mining textbooks, Bauer's *De re metallica*, was published, and became well known to Spaniards at home and in the Americas.[2] Although much of the sophisticated Saxon methods were not applicable to Mexican conditions, Bauer's cool realism influenced his readers for the better. For that matter, Spaniards of all ranks entertained great respect for German learning; one Karl Jedler was appointed administrator-general of homeland Spanish mining in 1594,[3] and German technicians were recruited in the later Bourbon period to serve in Mexico itself.[4]

Prospects were rich both in Mexico and Peru, areas of tilted-block mountain building with the associated profound acidic fissuring so helpful to profitable mineralization. After skimming off by war the patiently accumulated placer gold of the Aztecs and the Incas, the Spaniards promptly turned their attention to the lodes and placers whence it came. They never neglected to catechize any Indio they encountered regarding mineral indications nor to reward the bringer of interesting specimens. Even common soldiers quickly learned to distinguish at a glance the flat scales or nuggets of placer gold, the rusty honeycombed dinginess of gold-bearing quartz, and the characteristic appearance of surficial silver salts. Employing virtually all the methods previously outlined, the Spanish *gambusinos* rapidly prospected Mexico.

The conquistadors' first care was doubtless to annex the Indians'

[1] Alan Probst, "Bartolomé de Medina: The Patio Process and the Sixteenth Century Silver Crisis," *Journal of the West*, Vol. VIII (January, 1969), 90–124 (hereafter cited as "Bartolomé de Medina").

[2] Juan Mateo Manje, *Luz de tierra incognita* (trans. by Harry J. Karns), 88–89. Manje refers to Bauer as Barúa.

[3] Clement J. Motten, *Mexican Silver and the Enlightenment* (Philadelphia, 1950), 6.

[4] Walter Howe, *The Mining Guild of New Spain and Its Tribunal General, 1770–1821*, 306ff. (hereafter cited as *The Mining Guild of New Spain*).

placering grounds from which they had separated gold and, what was far more valuable from the Indian viewpoint, nephrite, or jadeite. Now placering is essentially the commercial separation by gravity of native (that is, reasonably pure) metal, or heavy metallic salts from naturally crushed gangue rock. It is conducted usually but not invariably along the banks and bars of watercourses and is a contender for the honor of being the most ancient method of obtaining metal. The particular metal was most probably gold, although silver and copper can be found in the same manner. Whether placering preceded or followed the discovery of fire smelting of ores is impossible to determine. The three basic methods of gravity separation practiced from earliest times were, in probable order of discovery, hand picking of nuggets, water separation, and air winnowing. All three methods have so much to recommend them that they are still in use.

Hand picking depends on the metallic glint of untarnishable gold, which attracts first the eye and then the hand—the experience of James Marshall at Sutter's mill is a good example of gold's attraction in this manner. Air winnowing was either borrowed from agriculture or lent to agriculture from placering—there is no telling; indeed, much of the machinery used for the two arts in primitive times was identical and interchangeable.

Water separation is a more complicated process, but can be traced far back. The legend of the Golden Fleece sought by late Bronze Age Greeks from the eastern Black Sea (traditionally from Colchis) has been interpreted in many ways. It is probable that early-day placer men of the Black Sea region (who incidentally claimed to be of part-Egyptian descent) anchored raw sheep fleece with cobbles to the beds of the fairly swift tributaries of the Phasis River. They then sank their pits near the water and transported the pay sand in baskets to the stream, where they tossed it by handfuls into the water above the wool. The gold dust promptly sank and was entangled both in the wool and in the gummy lanolin, while the much lighter sand was washed downstream by the rush of the water. (Low earthen dams with adjustable water gates were built just upstream to regulate the speed of the water.) After due time the fleeces were raised and washed clean in a large pot.[5] If the water in the pot was tinctured with the lye of wood ashes, the dust was more

[5] Aitchison, *A History of Metals*, I, 168.

effectively freed from the adhesive lanolin. It was then a simple matter to decant the wash water, dry the dust, puff it free from the silty residue, and either pack it in the usual goose quills or melt it down in a clay crucible for casting into bullion. For three thousand years this method was not greatly improved upon.

The Egyptians either invented or borrowed the stairstep progenitor of the sluice, but, mining in a region of the country devoid of wood, they carved it from stone and passed it on to the Greeks, who used it without alterations. Although the record is decidedly unclear, the Romans may have refined the stairstep separator into a true sluice or into what the Germans later called a buddle and the Spanish a planilla. The buddle, much praised by Bauer, was a smooth-surfaced, very slightly inclined table over whose face a thin sheet of water and placer sands slowly trickled. Its gentle washing carried away the light sands, leaving the heavier metallic residues at rest on the table surface by virtue of their superior specific gravity (gold is slightly more than nineteen times heavier, volume for volume, than water, whereas the average acidic gangue is little more than three to four times heavier). The buddle was too slow and delicate for the American miners, who preferred the sluice or riffle box, which had a steeper one-to-sixteen slant, a greater ton-per-hour capacity, and the ability to run untended. In Spanish times, the sluice was little more than a hollowed half tree trunk whose bottom was crossed with wooden or iron-shod baffles (riffle bars) in its upper reaches and on which a coarse woolen blanket or sheep fleece was secured toward the lower end.

Placering

The Spaniards worked *lavaderos*, or placers, by methods differing little in principle from later American concentration of hard-rock minerals. The fundamental tool was the batea, a shallowly conical washing bowl; the main production device was the sluice. Gravel banks were exploited in a peculiar fashion which crudely anticipated American hydraulic mining. Such a bank was cut into a series of stairstep terraces with a deep receiving ditch excavated at the foot. Dams and flumes were arranged so that a thin sheet of water cascaded downward over the entire length of the topmost terrace. Gangs of slaves then mounted the terraces, working them into a pulp with their feet and backing down as the earth was washed away. They continued to shuffle and retreat until the entire bank was trodden away and washed down into the ditch. The heavy deposit of gold and gravel in the ditch was then hand-picked to free it of cobbles, and the silt-gold residue panned out with a batea; mercury was added to float off the silt and black-iron sand which are the inevitable concomitants to placer cleanups.

Now and then placer men encountered exceptionally stiff clays which contained enough free gold to be worth working, even though clay was singularly aggravating to deal with, being too stiff and cohesive to be run through a sluice and too plastic to be crushed. Clay, usually the product of the weathering of the feldspar component of granitic rocks, was found both in placers and in mines; it was an unmitigated nuisance wherever encountered. It could be placered by puddling, or mixing the clay and a quantity of water in some sort of container, such as a barrel or a mortar boat, and stirring and raking until a more fluid gumbo was bailed away, and the remaining mess was panned or riffled. Though still practiced, puddling is not recommended except to persons who are enamored of hard labor and short commons or who, like early-day slaves, are in no position to refuse the work.

In Mexico mining began momentously with the location in 1543 of the Espíritu Santo Lode near Compostela, the capital of New Galicia, somewhat north and west of Mexico City. The discoverer was an Indian in the service of Doña Leonor de Arias (the impoverished widow of one of Nuño de Guzmán's officers), to whom he showed a specimen of the "ore."[6] It may be inferred that ore really meant a spray, or plaque, of native silver, for true surficial silver ore is usually an earthy green or beige mass (argentite or cerargerite) of crumbly texture whose only remarkable characteristic is its great weight. It is improbable that an Indio at that place and time would have been capable of associating such ore with any value, whereas native silver, being clearly metallic, would virtually demand investigation.

The strike set off a silver rush which continued through the next decade and during which were located the bonanza lodes of Zacatecas, Guanajuato, the Real del Monte at Pachuca, and San Luís Potosí.[7] The

[6] Trinidad García, *Los mineros Mexicanos*, 133ff.

[7] Marvin D. Bernstein, *The Mexican Mining Industry, 1890–1905*, 9 (hereafter cited as *The Mexican Mining Industry*).

gambusinos quickly found that these lodes were most commonly encountered in small hillocks, characteristically light brown in color and shaped like truncated cones. (In the American West such hillocks would be called blowouts, good examples being those at the Silver King and Iron King mines of Arizona.) The country rock of the hill was usually composed of a much-weathered porphyry or andesite, whose original purplish or greenish components had been leached away and replaced by the brown stain of iron originating from the oxidized silver lode within. The hillock's elevation above the mean surface was the product of the metamorphism which rendered the lode's casing rock slightly more resistant to erosion than was that of the surrounding countryside. The superficial resemblance to a miniature volcano is very deceptive to the untutored eye but is only coincidental.

The rush of 1543–53 produced about twenty-two million pesos of reported taxable silver, a sum much exceeding the value of all the silver seized during the Conquest itself.[8] Production was particularly economical and elementary, since the ore was mined by open-cast quarrying that required only simple hand tools and baskets that could be used by unskilled laborers. The silver chlorides were concentrated by hand sorting and gravity separation in water and were then reduced by direct smelting in charcoal-fired furnaces without much need for air blasting or fluxing. The smelter pigs were virtually pure silver and, in view of the European money famine that had prevailed since Roman times, commanded fantastic purchasing power. Anyone fortunate enough to come across such a lode required only charcoal, wrought iron, and Indian labor to become immoderately wealthy.

The Spaniard "denounced," or claimed, a gold or silver lode, to the local authorities. He had to promise faithfully to pay His Most Catholic Majesty the *quinto*, or one-fifth the gross bullion output, and was in turn usually lent some military assistance in recruiting a labor force from among the inhabitants of the region. Operations commenced along lines which would have been familiar to Herodotus. Pick and gad were adequate to break out the weathered chlorides and rotten quartz in the

[8] There is reason to suppose that not all production was declared and the royal taxes paid to the crown. Much of the surficial native silver could have been pocketed, or high-graded, by the miners themselves and much more quietly retained by the owners, as either bullion or plate.

oxidized zone above the water table. When that came to an end and tough quartz or porphyry was encountered, fire setting might be employed. This method involved building (with due attention to economical use of full and adequate ventilation) a hot, prolonged fire against the face of the heading, then dashing cold water against the glowing face to break it out by thermal stress. Innumerable writers have held that vinegar was often employed in place of water, implying that, being a weak acid, it would profitably attack basic (alkaline) rock such as limestone or Alpine dolomite. Both quartz and porphyry are acid rock, however, and thus are immune to vinegar. Besides, there is some reason to think that the vinegar story was made up from whole cloth.[9] The fire served additionally to pulverize the gangue in preparation for milling, drive off sulphides (the sulphurous reek of fire setting was notorious), and sweat free-milling mineral particles therein into larger and hence more readily recoverable droplets (Fig. 10).

Fire setting was considerably more expensive than "cold mining" by human muscles alone. It required enormous quantities of cordwood, which had to be cut from local forests (where available). It made the workings more or less uninhabitable for several days, filling them with heat and deadly gases, which had to be conscientiously evacuated with fans, shaft-head windsails, and similar devices. In large ore bodies of very high tenor fire setting was not without risk to the workings themselves, for rich sulphide ores have been known to catch fire (hence the name pyrites) and burn in place underground for an indefinite period of time. In the 1920's such an ore fire resulted from an accident in the United Verde workings at Jerome, Arizona, and in the end forced a complete revamping of the operation from drift mining to open-pit excavation. The heat of these fires also often altered the ore so as to make it singularly difficult to concentrate and mill.

When fuel was in short supply, as it usually was in the mining districts of Hispanic America, cold mining alone was employed. The chief if not sole tool of the Indian miner was a long iron crowbar or gad weighing twenty to thirty pounds. It served as pick, drill, and moil and accounts for the name commonly given an Indian miner—*barretero*.

[9] Gösta Sandström, *Tunnels*, 28–30, vigorously contends that the distillation of wood occurring in fire setting released natural acetates whose vinegary smell deceived bystanders.

Fig. 10. Fire setting in a Renaissance mine in Saxony. *A*, kindled logs; *B*, sticks shaved down fan-shaped; *C*, tunnel. From Georgius Agricola [Georg Bauer], *De re metallica* (1555).

Later on in the United States, the bar was sometimes called a jumper and was often sprung from a green sapling to lessen the effort required. The ore was collected in fiber baskets of up to 350 pounds' capacity, secured to the forehead by tumplines, and carried out by *tenateros*, or porters, climbing notched *boca-mina* tree trunks called chicken ladders, often ascending hundreds of feet (Fig. 11).[10]

[10] Alexander von Humboldt, *Political Essay on the Kingdom of New Spain*, Book IV [Vol. III], 234–35 (hereafter cited as *Political Essay*). Provision for earlier chicken ladders is to be seen in the shafts of the mines of Laurium (cf. Sandström, *Tunnels*, 36). The word *tenatero* comes from *tenate*, the fiber sack used for ore haulage.

Fig. 11. A Mexican *barretero* with *tenatero* descending chicken ladder. From Louis Simonin, *La vie souterrane* (1867).

As far as the sinking of shafts or the driving of headings went, engineering practice appears to have been regressive even by Roman standards. The miners drifted, or directly followed the ore lead, making no attempt to utilize the force of gravity in moving ore and waste, to anticipate geological developments, or to find shortcuts to new ore masses.[11] The cross section of the heading was a coyote hole the width of the lode plus barely enough room for the miners to jounce their bars, no matter how great their inconvenience or discomfort.[12] When the lode was small, the heading might actually have a keyhole cross section, wider for the lode (and the miners' shoulders) and much narrower for the feet. If the lode pinched out, went into *borrasca*, flooded, or in any other way failed to remain attractive, the mine was promptly abandoned.[13] Only reasonably self-supporting hard rock could be worked, though masonry pillars were occasionally installed in large chambers. Methods of pumping more effective than direct bailing with a bucket hoisted by a whim, or *malacate* (Plate 1), were entirely out of the question.[14] There were some very good reasons for these seemingly inadequate procedures, and many an abandoned Spanish-period mine was later revived and made to pay well by the introduction of more efficient scientific mining methods—but for that matter, virtually every mine or placer anywhere has been reworked time and again as techniques grew more refined or markets improved.

Before 1550 slavery pure and simple was employed by the conquistadors in the shallow mines and placers. So little had matters changed in three thousand years that Diodorus of Sicily's description of life in the mines of Egypt would serve with hardly any editing for life in the Mexican workings in the earliest Spanish period:

At the extremity of Egypt and in the contiguous territory of both Arabia

[11] Humboldt, *Political Essay*, Book IV, 238.

[12] Motten, *Mexican Silver*, 14ff.

[13] A pinched-out lode was one which an earth fault had crossed; subsequent ground movement had then displaced the farther side of the lode to such an extent that its recovery was difficult without the geological theory of a later day. A *borrasca* lode was one whose mineral content had suddenly thinned greatly or even vanished for a variety of causes.

[14] A whim was the animal-powered equivalent of a ship's capstan, the drum or spool axis being vertical, whereas the axis of a hand-operated windlass was horizontal.

and Ethiopia[15] there lies a region which contains many large gold mines, where the gold is secured in great quantities with much suffering and at great expense. For the earth is naturally black[16] and contains seams and veins of a marble [quartz] which is usually white. . . . The kings of Egypt gather together and condemn to the mining of the gold such as have been found guilty of some crime and captives of war, . . . by this means not only inflicting punishment upon those found guilty but also securing at the same time great revenues from their labours. . . . The gold-bearing earth which is hardest they first burn with a hot fire, and when they have crumbled it in this way they continue the working of it by hand; and the soft rock which can yield to moderate effort is crushed with a sledge by myriads of unfortunate wretches. And the entire operations are in charge of a skilled worker who distinguishes the stone and points it out to the labourers; and of those who are assigned to this unfortunate task the physically strongest break the quartz-rock with iron hammers, applying no skill to the task, but only force, and cutting tunnels through the stone, not in a straight line but wherever the seam of gleaming rock may lead. Now these men, working in darkness as they do because of the bending and winding of the passages, carry lamps bound on their foreheads. . . .

The boys who have not yet come to maturity entering through the tunnels into the galleries formed by the removal of the rock, laboriously gather up the rock as it is cast down piece by piece and carry it out into the open to the place outside the entrance. Then those who are over thirty years of age take this quarried stone from them and with iron pestles pound a specified amount of it in stone mortars, until they have worked it down to the size of a bean. Thereupon the women and older men receive from them the rock of this size and cast it into mills of which a number stand there in a row, and taking their place in groups of two or three at the spoke or handle of each mill[17] they grind it until they have worked down the amount given them to the consistency of the finest flour. . . .

[15] These mines were west of the Red Sea sector of the Great Rift Valley, their outcroppings appearing on a series of plateaus sloping westward toward the Nile Valley. A similar series of auriferous formations lies east of the Red Sea at the tip of the Arabian Peninsula.

[16] This black rock was schist, a greatly metamorphosed sediment whose presence is considered a favorable indication of possible mineralization (Walter H. Emmons, *Gold Deposits of the World, with a Section on Prospecting*, 384–88; hereafter cited as *Gold Deposits of the World*).

[17] It is a pity that Agatharchides, whose lost work on geography Diodorus was quoting, had not seen fit to describe these mills in greater detail, since they may have been arrastras, Chilean wheels, or even Roman mills. Rickard, *History of Metals*, says they were Roman mills, but his reasoning is weak.

> In the last steps the skilled workmen receive the stone which has been ground to powder and . . . rub [it] upon a broad board which is slightly inclined, pouring water over it all the while; whereupon the earthy matter in it, melted away by the action of the water, runs down the inclined board, while that which contains the gold remains on the wood because of its weight.[18]

Subsequently the board was cleaned with sponges, and the dust was refined by being melted with lead, salt, and an oxidizing agent and was cast into bullion.

The Greek observers were appalled by conditions in the Egyptian mines, forgetting or ignoring the fact that the slave miners of their own Laurium were treated as badly, if not worse. J. Frank Dobie described the unpleasant working conditions in the Mexican mines and recounted legends of ghosts of miners who perpetually lamented their lot, pushed ore cars about, and induced rockfalls. It should be noted that the legends ascribing much of this mining to the Jesuit order were emphatically denied by Hubert Howe Bancroft.[19] By Alexander von Humboldt's time matters had been improved considerably; Humboldt noted that the Indian miner was a free man, was comparatively well paid, and had even learned to make off with rich ore when the opportunity presented itself.[20]

Crushing

Having been hoisted or carried from the mine, the ore was dressed by being exposed to the air, broken coarsely with sledges, and heaped in a yard where it was cobbed, or sorted.[21] Waste was thrown away, high-grade pieces were segregated for smelting, and middling-tenor ore was held for patio treatment (Fig. 12). Fuel supply permitting, it was next roasted or calcined to further break up the gangue and to expel arsenides, sulphides, and, in the case of some gold ores, tellurides.[22]

[18] Diodorus Siculus, *Diodorus of Sicily* (trans. by C. H. Oldfather), III, 12–14.

[19] Cf. J. Frank Dobie's *Apache Gold and Yaqui Silver* and *Coronado's Children* for specimens of folklore. The mining activities of the Jesuits were denied by Bancroft in *History of Arizona and New Mexico, 1530–1888*, 362.

[20] Humboldt, *Political Essay*, Book IV, 247–48.

[21] Ore was cobbed, or physically upgraded, on the ore platform or ore shack by workers who knocked away with an iron hammer on a rough anvil valueless pieces of gangue rock.

[22] Dry-masonry edifices obviously used in processes involving combustion are

Fig. 12. Apache prisoners cobbing ore in northern Mexico. From Louis Simonin, *La vie souterrane* (1867).

Fig. 13. A primitive Mexican ore crusher. From Louis Simonin, *La vie souterrane* (1867).

The ore then went to the grinder or crusher. There was quite a variety of such implements. The most primitive was a boulder lashed into a substantial wooden fork, supported at the balance point upon another fork. A worker sawed up and down on the handle, causing the boulder to thump into the shallowly hollowed stone mortar which held the ore (Fig. 13). Slightly better was the pillow block (later called the

frequently found about abandoned mine workings in the Southwest. Dilapidated constructions are difficult to identify, but smelter stacks tend to be relatively small and thin and are sometimes built into a hillside. Charcoal kilns and ore roasters are shaped somewhat like beehives. A kiln might well be the more tightly constructed, so as to exclude unwanted air, and would probably be sited farther from the shaft collar or portal than a roaster would be. A series of such beehives in line suggests kilns, inasmuch as in most cases a single roaster would be sufficient to handle the production of a primitive mine.

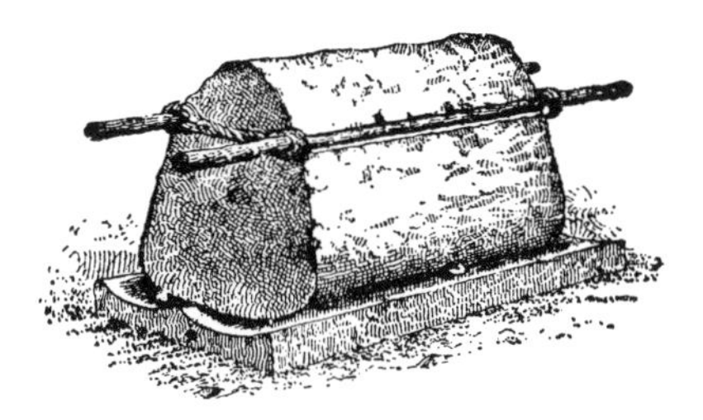

Fig. 14. The pillow-block, or Korean, ore crusher. From Joseph C. F. Johnson, *Getting Gold* (1917).

Korean block), whose upper stone's surface was slightly curved laterally. Ore was scattered in the space between the upper stone and the nether stone on which it rested, and the pillow block was rocked laterally by two men who pulled and pushed on the wooden handles with which it was furnished (Fig. 14). Slightly better still was the *trapiche*, or seesaw grinder, which utilized the men's weight and leg muscles as they seesawed up and down on a plank lashed to the spherical grinding boulder, which rocked as well as rotated within the cup-shaped stone mortar.

The chief and ultimately permanent favorite crusher of the Spaniard (and the shoestring American miner to come) was the arrastra,[23] whose origin was the threshing floor of the ancient Middle East. It was mentioned by Pliny and was undoubtedly taken to Iberia by the Phoenicians. The device consisted of a circular floor of andesite blocks, laid tightly upon a subflooring of well-puddled clay, with the whole laid level or at a slight angle to the horizontal.[24] A low stone coping, pierced by an outlet scupper at the low end, ran around the floor. In the center was set a substantial wooden post on a stone pillar-pivot. One or two beams were set horizontally on the pivot, and from the beams were loosely hung pairs or even quads of 150- to 200-pound drag stones of the same andesitic material. The drag stones were beveled slightly on the lower

[23] From *arrastrar*, meaning "to drag." The same device was called kollergang by the Germans. The remains of such a device, dating from the fifth century B.C. have been identified at Laurium (Sandström, *Tunnels*, 38 note 4).

[24] Andesite is any fine-grained igneous rock whose acidity is intermediate between granite and basic. For arrastra-building purposes, however, what was really desired was a stone which would burr rather than polish with continued use (*cf.* Theodoro Köhnicke, "Inexpensive Home-made 20-Ton Mill," *Mining and Scientific Press*, Vol. XCVII [August 8, 1908], 185–86).

leading edge and were suspended thereby so as to "nip" the ore. The end of one beam extended over the coping and was provided with a harness for horse or mule. When used as a simple grinder, the arrastra was shallowly filled with crushed, dressed ore, the animal, blindfolded, was harnessed to the beam end, and a trickle of water was set going into the ore. As the animal promenaded about the arrastra, the drag stones ground the ore to a thin pulp or slime, which dribbled out the screen (in early times a pierced oxhide) in the scupper, whence it was led away for further treatment (Fig. 15).

The arrastra (further corrupted to "raster" or "rastus" in semiliterate memoirs) was not the best invention in the world, but it had two important virtues: it could be cheaply erected almost anywhere by anybody, and it used only animal power (though it has been reported that avaricious miners of the American period were prone to put their Indian helpmates to work at the poles when the horse or mule needed a rest).

When amalgamation came into general use, the arrastra technique

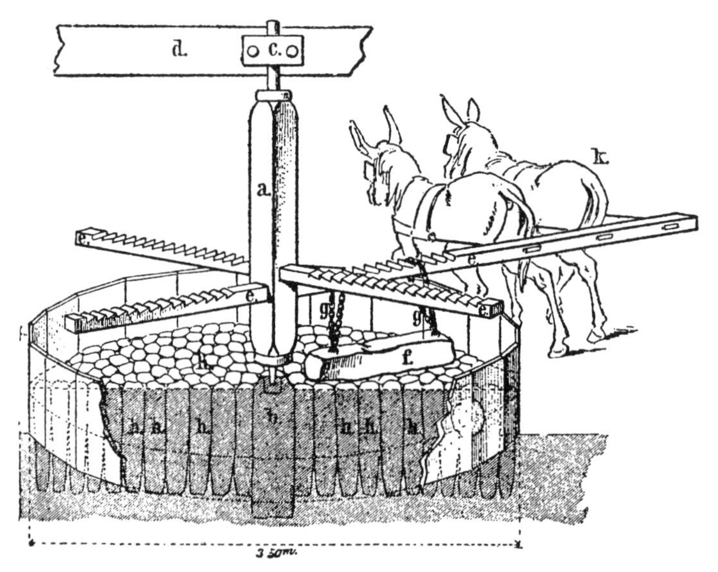

Fig. 15. An arrastra, showing one drag stone. From Thomas A. Rickard, *Journeys of Observation* (1907).

was suitably modified to take advantage of mercury's ability to seize upon and unite with gold and silver. As in the earlier process, the arrastra was charged with a layer of dressed ore, which was then wet-ground to the size of middling sands. At this point mercury was sprinkled on the floor through a cloth tied about the neck of a jug or a beer bottle, in proportion to one ounce to each estimated ounce of free-milling gold in the charge. Grinding was recommenced and continued until the batch was slimed. A little time was allowed for the amalgam droplets to settle through the floor blocks; then the scupper was opened and the exhausted tailing sluiced away, its destination being a riffle box. The operator now pulled up and carefully washed the flooring blocks in a vat, then with an iron or horn spatula collected the amalgam from the surface of the clay pan beneath. The accumulation was washed clean, and more mercury was added to make the lump sufficiently fluid to float off the "black sand" (iron particles). The cleaned amalgam was then retorted or distilled to recover the gold-silver sponge from the mercury.[25]

In this modification the arrastra became a sophisticated and reasonably satisfactory triple-function mill, acting in succession as wet grinder, amalgamator, and separator. In economy of construction it could hardly be improved upon, and it has been employed to the present by small-time operators without any modification except substitution of concrete for clay. While good for the miner, this fact presents difficulties for the archaeologist, for it is virtually impossible to determine by physical evidence alone whether the remains of any given arrastra date from Spanish, Mexican, or early American times. Also, since it was usually profitable to dismantle a time-expired arrastra and riffle the ground beneath it, into which amalgam had seeped through the clay pan, intact specimens are scarce (Plate 2).[26]

The Chilean wheel, sometimes called the edge runner or Chile mill by Americans, was a superior crusher capable of being powered by a water mill. Using the basic stone platform and coping of the arrastra,

[25] Humboldt, *Political Essay*, Book IV, 258; *Mining and Scientific Press*, Vol. XCVII (November 7, 1908), 627.

[26] It is difficult to exaggerate the evasiveness of mercury. With the recent boom in mercury prices it has been found profitable to excavate and rotary-kiln ground lying one hundred feet below the old roasting ovens at New Almaden, California, so as to recover metallic mercury escaped from the retorts in the 1850's.

it substituted for the drag stones one or two large stone wheels set on edge and revolving about an axle set horizontally on the central post. Since the identical mill was used in Palestine in biblical times for the preliminary crushing of olives and oilseeds, its ascription to Chile has nothing to do with its invention. The wheel arrangement eliminated a great deal of wasteful friction and nipped the ore just right, but, being difficult to dismantle often, was probably used almost exclusively for preliminary crushing of dry ore. Chilean wheels had the additional disadvantage that their cutting and shaping required the services of an experienced mason. By Colorado mining days American manufacturers were advertising what amounted to knockdown Chilean wheels of cast iron, but it is doubtful that they were widely used (Plate 3).

The last important Spanish machine to be developed was the *maza*, a greatly enlarged mortar and pestle consisting of a square wooden pillar stamp working up and down in a square stone mortar which closely fit the lower end. When it was powered by a water mill, it was referred to as an *ingenio*. The system was not a Spanish invention; stamp mills were employed in Cornwall, and Bauer presented in *De re metallica* a clear illustration of multiple gangs of such stamps, each gang running off its own overshot wheel (Fig. 16). The *maza* had its shortcomings, not the least of which was that it had to be stopped at the end of a batch run, the mortar awkwardly cleaned by hand, and a new ore charge shoveled in for the next run. The *maza* was evidently a progenitor of the "California" gravity stamp mill, and was critical to the *cazo* silver-reduction process; yet the lack of waterhead in arid Mexico seems to have forced the Spaniards to rely chiefly upon the animal-powered arrastra and Chilean wheel (Fig. 17).[27]

The Patio Process

By 1553 the great Mexican silver bonanza had begun to run into serious difficulties. With the stoping out of the oxidized zone of naturally concentrated and readily smelted ore, a number of grave problems automatically arose. The country rock changed from sound andesite to a much-fractured shale which was permeated by the permanent water table. In order to continue sinking on the ore, pumping or bailing was necessary. The quartz gangue containing the silver became much harder

[27] Humboldt, *Political Essay*, Book IV, 257.

Fig. 16. A Saxon stamp mill driven by water power. *A*, first machine; *B*, its stamps; *C*, its mortar box; *D*, second machine; *E*, its stamps; *F*, its mortar box; *G*, third machine; *H*, its stamps; *I*, its mortar box; *K*, fourth machine; *L*, its stamps; *M*, its mortar box. From Georgius Agricola [Georg Bauer], *De re metallica* (1555).

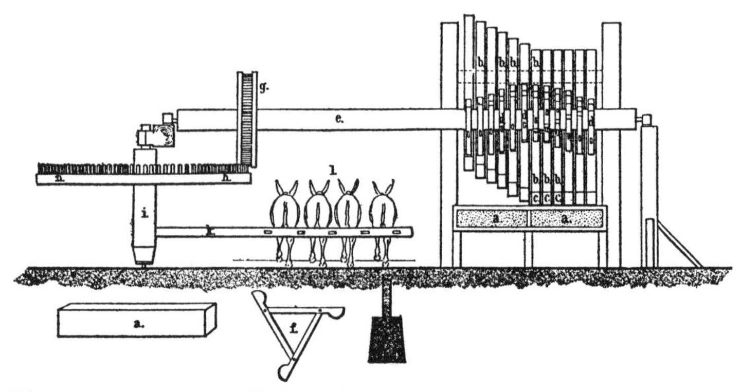

Fig. 17. A ten-stamp mill driven by animal power. *a*, battery; *b*, stamps; *c*, shoes; *d*, lifter pegs; *e*, drive shaft; *f*, side view of wooden cam; *g*, vertical crown gear; *h*, horizontal crown gear; *i*, vertical drive shaft; *k*, harness pole; *l*, mules. From Thomas A. Rickard, *Journeys of Observation* (1907).

in texture, the subsurface water having protected it from weathering and decomposition. The tenor of the ore declined abruptly, since the water protection had prevented the atmosphere from naturally concentrating its silver values. Worst of all, the nature of the ore changed from chlorides to sulphurets of silver. In the sixteenth century sulphurets could not be reduced by any means other than a complicated smelting process which was costly in flux and fuel and which the Mexican mine operators were incapable of imitating for want of the appropriate materials and techniques. Collapse of the entire industry seemed imminent; indeed, in modern times, wherever ore persists in depth, striking the water table automatically ends bonanza and necessitates complete and costly revision of the whole operation. Possibly the only saving factor was that the Mexican ore did persist as silver; much more often than not, a high-grade silver chloride ore changes at this critical point to a copper-sulphide *borrasca* ore of very low tenor which the miners of the period were unable to work.

Rescue came with remarkable speed in the person of Bartolomé de Medina, one of the company of dedicated amateurs who occasionally revolutionize technology. It is fair to compare Medina in silver to Bessemer in steel. As a merchant in Spain he learned from one Lorenzo, a German, that mercury and common salt would reduce silver chlorides.

He determined to adapt this process to the reduction of silver sulphurets and moved to Pachuca, Mexico, where he engaged in years of experiments with that district's sulphide ores. In 1554 he added magistral, obtained by roasting copper-iron outcrop ore to form an impure blue vitriol, or bluestone (copper-iron sulphate). It proved the key to his formula, the patio process, which was one of the first and greatest triumphs of empirical industrial chemistry, long predating equivalent developments in northwestern Europe. The process required virtually no fuel or flux and no great amount of mechanical power, it was ideally suited to a semiarid land, and it employed no ingredients (save salt and mercury) which were not relatively near to hand. As is so often the case, Medina died in near penury, having himself made hardly a centavo from a technical innovation which would dominate silver milling until the 1890's.[28]

The process, as conducted for the following three hundred years, began with crushing the silver ore under a Chilean wheel to the size of kernels of corn. The ore was next reduced to middling sands in an arrastra and screened by being passed through a pierced ox hide. The sands were then spread in a thin layer over a stone platform, or patio. Under the supervision of the *azoguero*, or "mercurist" (an experienced master workman whose badge of office was a small gourd from which he sprinkled in his reagents), large quantities of salt and copper sulphate were added and thoroughly worked in. After some days of mixing and trampling, originally done by human feet (in 1673, Juan Comejo, of Peru, sensibly substituted those of herds of mules), mercury was sprinkled on and likewise worked into the *lama*, or mix. On the basis of the *azoguero*'s judgment of samples taken and inspected at intervals, the *lama* was spread thinly and uniformly about the patio for exposure to heat and sunlight. It was subsequently raked and heaped into a *torta*, or mound, where it lay until another redistribution. The mix changed gradually to a gray color, and at the end of fifteen to forty-five days was judged to be sufficiently seasoned. It was agitated in water in windmill-powered vats to wash out the earthy gangue, and the silver amalgam remaining was collected for retorting (Plate 4).[29]

[28] Probert, "Bartolomé de Medina," *Journal of the West*, Vol. VIII (January, 1969), 1.

[29] Albert M. Gilliam, *Travels over the Table Lands and Cordilleras of Mexico*, 258–61.

The *azoguero* practiced a great deal of mumbo-jumbo, no doubt intended to fortify his institutional standing. If he decided that the *lama* was too "hot," he added lime to "cool" it; if too "cool," he sprinkled more bluestone on it to heat it up—neither step contributing one whit to the effectiveness of the process. He did help earn his pay by his estimation of the amount of mercury required, for this rare metal was a most expensive crown monopoly, and it would never do to be unthrifty with a metal which was nearly as costly as silver itself.[30] The Mexican silver patios consumed five to six thousand quintals (flasks) of mercury annually, most of which was produced from the great Spanish deposits at Almaden, which had been worked since classical times. Instead of permitting a free market, the crown maintained a monopoly on the metal, selling it through Cajas Reales (Royal Offices) at arbitrarily fixed prices which varied from $60 to $110 a quintal but which were calculated to yield a profit of not less than 100 per cent. Development of known Mexican cinnabar deposits was prohibited,[31] although it was grudgingly permitted that Austrian mercury from Idrija or Peruvian quicksilver from Huancavelica might be sold at the inflated price when the Almaden product was in short supply.[32]

Retorting employed the moderate heat of burning charcoal to separate the mercury from the silver retort sponge. For this process the Mexicans employed a large bronze bell, the *capellina*, named after its developer, Juan Capellín, who may have got the idea from an unsatisfactory, small-scale, and more complex mercury-recovery structure depicted in *De re metallica*.[33] Although the massive bronze bell required a chain fall and a gang of men to manipulate it, it had the virtues of simplicity, effectiveness, and safety in protecting the workers from the deadly fumes of mercury vapor (Fig. 18). It was another excellent illustration of Mexican ingenuity in adapting complicated procedures to the limited experience of Indian workers, as was the crushing equipment, which required scarcely any iron and only minimal amounts of wood, relying instead upon the plentiful stone and leather.

Most visitors to the Mexican silver belt commented at greater or lesser

[30] Humboldt, *Political Essay*, Book IV, 90–124.
[31] Howe, *The Mining Guild of New Spain*, 11–12.
[32] Humboldt, *Political Essay*, Book IV, 280–84.
[33] Motten, *Mexican Silver*, 27; Bauer, *De re metallica*, 427ff.

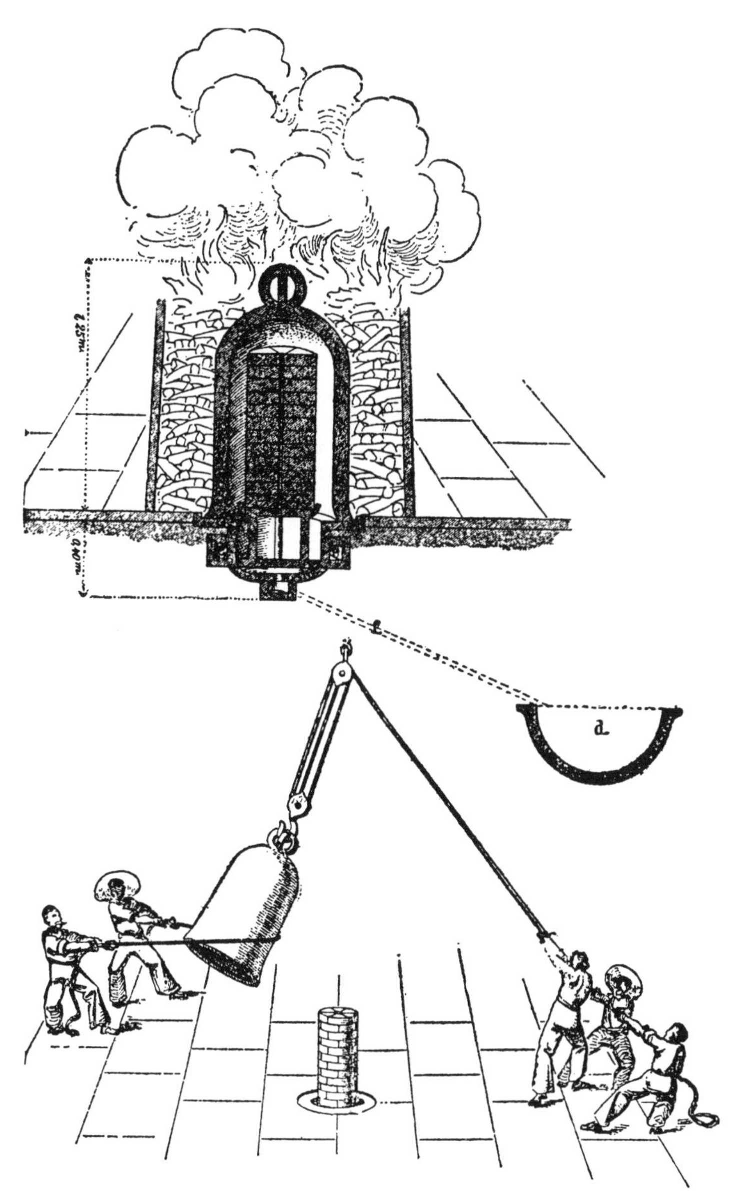

Fig. 18. Mexican retorting of amalgam with *capellina*. From Thomas A. Rickard, *Journeys of Observation* (1907).

length upon the patio process, finding it fascinating and quite unlike anything they had ever seen before. Possibly the best and most detailed of the accounts is that of Humboldt, and yet the great scientist assumed an air of patronizing criticism which, however well founded in the latest scientific discoveries, tended to ignore local conditions and to assume that what was good for Saxony must necessarily be good for Mexico.[34] A half century later the Freiburg-trained engineer Herman Ehrenberg, better acquainted with the problems of arid-lands metallurgy under primitive conditions, was still recommending the patio process as the one most ideally suited to country which lacked skilled workmen and the wherewithal of mechanical power.[35]

Although Humboldt had more than a fair idea about the workings of the chemistry involved, the true action of the patio process was not understood by millmen as late as Comstock times. The theory generally prevailed that it was an alchemical "digestion" accomplished by the salt and copper sulphate, followed by simple amalgamation of the silver with the mercury. On the contrary, a fairly complex set of reactions was involved. The common salt and copper sulphate reacted to form cupric and cuprous chloride. They in turn reacted with the silver sulphurets to form silver chloride. The silver chloride was dissolved by the brine present and was reduced by the mercury to be precipitated in metallic form which amalgamated with the remaining quicksilver. In the reduction some of the mercury was transformed to soluble chlorides and bichlorides of mercury, which were washed away in the final separation process. Thus a certain amount of quicksilver was dissolved and lost, to the great puzzlement of those who nursed the digestion theory.[36] William Wright, the Washoe chronicler, admitted mystification at the large-scale disappearance of mercury in the process but did not offer to account for it except to say that it went "up the flue or down the flume" but that no man knew whither.[37]

Wright also observed that the mercury mysteriously tended to reappear in the presence of iron (which, of course, reduced it back to the

[34] Humboldt, *Political Essay*, Book IV, 257–80.

[35] [Herman Erhrenberg], "The Reduction of Silver from Its Ores," *Weekly Arizonian* (Tubac, Arizona), Vol. I, No. 22 (July 21, 1859).

[36] D. M. Liddell, *Handbook of Nonferrous Metallurgy*, II, 525; J. B. Mannix, *Mines and Their Story*, 175.

[37] Wright, *The Big Bonanza*, 96–98.

metallic form). For instance, nailheads in flumes became coated with quicksilver to a degree which made its salvage economical.[38] Likewise, distinct deposits or pools of mercury might be found in pockets of the stream bedrock into which flume water from the mills was drained. No effort at all was made to take advantage of this fact, and the losses in Mexico alone amounted to 90,000 to 160,000 pounds annually, or about 1.25 pounds per ton of ore treated.

El Sistema del Rato

The Mexican canon of mining and silver metallurgy gained the name *el sistema del rato*, which is translated literally but awkwardly as "the system of the moment" (its true meaning is better implied by the phrase "pragmatic system" or "empirical system"). The term persisted from the Conquest to the accession of Porfirio Díaz in 1877, with his *científico* program of opening Mexico to foreign mining experts and corporations. In the days of Díaz, American and British mining engineers, scornful of all things indigenous, willfully mistranslated the phrase as "rat-hole mining."[39] They observed that a mine of any long standing was a labyrinth of twisting, dipping passages in which an unguided visitor could quickly become hopelessly lost. No surveys had been made; no maps had been drawn. Necessary later works such as hoist and pump shafts were frequently ineffective, having been sunk by guess and by God, often representing a total waste of time and money. Blasting (introduced in the Bourbon period) was practiced enthusiastically but ineffectively. The movement of ore and waste would have made an Athenian weep with chagrin.

In his early-nineteenth-century investigations Humboldt excoriated the workings as wretched and of medieval barbarity, comparing them to "ill-constructed buildings where, to pass from one adjoining room to another, we must go round the whole house."[40] These strictures were not universally true. Some of the greatest mines, such as La Valenciana, were laid out and their access works constructed with monumental, even lavish, attention to architectural appearance. But for the most part

[38] *Ibid.*, 98.

[39] Kenneth M. Johnson, in *The New Almaden Quicksilver Mine*, gives an excellent description of rat-hole mining in the Mexican manner. See also Motten, *Mexican Silver*, 18–19.

[40] Humboldt, *Political Essay*, Book IV, 219, 232, 237.

the drifts and headings resembled the burrowings of termites. Thousands of horses and mules were worked to death at the *malacates*, the "blood hoists."

There were some very sound historical reasons behind this apparently haphazard and wasteful approach. First, it was not until 1791 (150 years after the discovery of the silver lodes) that Abraham Gottlob Werner published his *New Theory About the Origin of Lodes*, which for the first time placed mining geology on the correct theoretical basis. Within three years Andrés del Río brought Werner's ideas to Mexico.[41] Until the appearance of Werner's book no miner anywhere had been able to do more than drift or sink upon ore, and the peculiar structure of the Mexican deposits—totally unlike the relatively simple silver-lead deposits of Saxony described by Bauer—virtually demanded rat-hole burrowing. The basement rock of Mexico was andesite, a firm, fine-grained granitic flow. It was overlaid in considerable depth by the brittle, greatly fractured and fragmented shale, which was in turn capped by the surficial andesitic flows. At a later time the quartz and silver of the ore was injected from below by the rise of magmatic fluids and vapors. Upon penetrating the shale, these magmatic emanations were free to follow its numberless faults and fissures, depositing the ore therein, so that the resulting lodes came to resemble the complex twists and turns of yarn thoroughly tangled by a playful kitten.[42] Before Werner no geologist on earth could have worked these deposits by any means other than the one employed by the "backward" Spaniards.

Drifting after the ore was not only necessitated by the want of theoretical knowledge but also complicated by the absence of mine surveyors.[43] Mine surveying is one of the most recondite and demanding of engineering arts, and it is hardly too much to say that celestial navigation or topographical surveying is virtually child's play in comparison. The art must be taught by intensive formal training in higher mathematics, followed by an equally intensive practicum. That such education was not available is suggested by the fact that when Viceroy Luís de Velasco established the Royal and Pontifical University of Mexico in 1555, the university seems to have included little mathematics beyond plane

[41] Motten, *Mexican Silver*, 19.
[42] Rickard, *Journeys of Observation*, 36–40, 43–49.
[43] Humboldt, *Political Essay*, Book IV, 238.

geometry in its curriculum.[44] Yet in mining a bad start implies a bad end, for accurate surveying must begin before ground is broken, while one of the most maddening tasks an engineer can face is that of surveying and mapping a mine which has been worked by empirical drifting for any length of time. Likewise, later works cannot be well constructed, as instanced by the ineffective pump shafts which had to be sunk without sound information about over-all mine depth or existing inclination of the drifts—data essential to arranging good natural drainage to the sump of the shaft. Hoisting raises and winzes would be equally ineffective. Indeed, any rational exploration and development begins with the survey and its map.

In retrospect, however, the Spaniards cannot be too severely criticized. The state of the art was in infancy everywhere from 1550 to 1650, and the mine-surveying methods outlined by Bauer were effective only for lodes much shallower and far less complex in structure than those the Spaniards worked.[45] It has been said in all truth that "unsupervised native workings have made uneconomical the recovery of tons of ore otherwise commercially valuable."[46] But it is equally true that the only alternative at that time would have been to do no work at all.

Critics in the Díaz period also failed to consider the ineffectiveness of blasting techniques throughout the colonial period. When a mine is in lean ore (and the silver sulphurets below the water table were not remarkably rich on an ounce-per-ton basis),[47] the most profitable course appears to be to mine only this ore, breaking out and hoisting as little waste as possible. In any case, untrained mine superintendents preferred to minimize their expenditures on exploration and the construction of access works. It so happened that this "development work" was usually much higher in cubic-yardage cost (and to no immediate return) than was the profitable stoping of ore, owing to the peculiar fact that

[44] Irma Wilson, in *Mexico: A Century of Educational Thought*, 23–24, indicated that in the scholastic bachelor of arts curriculum offered in the University of Mexico, no mathematics more advanced than Euclidean geometry was taught.

[45] Bauer, in *De re metallica*, 129–48, suggested surveying limited to the use of analogous triangles, charting by compass and plane table or by compass and surveyor's line to lay out on the surface the course of a drift directly beneath. Though that method would do well enough in mines of Saxony, it would break down at once when attempted in the enormously deep and extensive Mexican lodes.

[46] Bernstein, *The Mexican Mining Industry*, 27.

[47] *Ibid.*

Fig. 19. Double-jack drilling of a down hole. From Louis Simonin, *La vie souterrane* (1867).

ores are often much more friable and easily broken out by hand than is the adamant country rock which encases them. The shortcut of blasting appears to have been unavailable before about 1770, another reason for the necessity of rat-hole mining.

Blasting with gunpowder was a Saxon development, and was probably introduced by German technicians to Mexico in the time of the

Bourbon reforms. For a long time it was handicapped by a variety of impediments. Effective blasting requires high-quality steel drills, good drilling practices, and reliable safety fuse. It is doubtful whether good steel was available in any quantity in Mexico at acceptable prices. Moreover, drilling practices were inefficient, and safety fuse had not yet been invented. Nevertheless, the Spaniard of the Bourbon period made the attempt, and perhaps on a greater scale than anywhere else in the world, since by 1810 as much as 14,000 quintals of gunpowder were being imported annually for the mines. Most of it was wasted, since for want of safety fuse the *pegadores*, or blasters, dared not shoot rounds of holes (multiple, nearly simultaneous blasts) but had to restrict themselves to putting down and shooting out one hole at a time (Fig. 19). To compensate, they mistakenly drilled their holes too deep, thinking that the increased powder capacity thereof would offset the bad characteristics of a single shot.[48] They were wrong, but the technical means to put matters right would not appear until 1831.

A success as clear-cut as Medina's patio process was that of the Spanish mine operators in dealing with their Indian miners (Fig. 20). The Indios commenced to mine in a deplorable condition of slavery or such an approximation of slavery (the *encomienda* system) as to make no substantive difference. There was no a priori reason to suppose that their miserable lot would improve; on the contrary, there was every such reason to suppose that it would decline, if possible, to the brutal degradation mentioned by such classical writers as Plutarch and Agatharchides. If the church offered to intervene, a conspiracy of proconsular viceroys and greedy mine operators could easily thwart religious humanitarianism. On the contrary, however, virtually the first act of Viceroy Luís de Velasco upon assuming office in 1550 was to insist upon emancipation of the mine slaves.[49] Thereafter the condition of the Indian miners was systematically and substantially elevated until Humboldt could report, with a slight degree of exaggeration, that they were the best-rewarded miners in the world.[50] All nineteenth-century accounts agree that the

[48] Humboldt, *Political Essay*, Book IV, 234.

[49] Silvio Zavala, *Los esclavos Indios en Nueva Espana*, 123–24.

[50] Humboldt, *Political Essay*, Book IV, 246–47.

Fig. 20. Chilean *tenateros*. From Louis Simonin, *La vie souterrane* (1867).

Mexican miners were courageous, enterprising, and, within the limits of their gross illiteracy, remarkably ingenious.[51]

Any explanation of their methods which claims that they were influenced by the foreign technicians may be dismissed. Neither the Saxon nor the later Cornish experts appeared able to convey many of their technical refinements to the Mexicans, who seemed content to do things the traditional way.[52] That being the case, how much less could the foreigners affect what amounted to a "closed" society and culture? The most tenable conclusion is that the crown, the church, and the mine owners unanimously agreed that it was both good Christianity and good business to adopt what was, for the age, an enlightened labor policy. Unlike the American mining frontier, which remained wedded to the status quo in wages and working conditions,[53] Mexico stands out as a remarkably successful example of experimentation in social improvement. The subject is deserving of a separate study, but even the brief above serves to refute the "black legend" of Spanish brutality and cruelty toward the Indians.[54] For that matter, the conditions of underground laborers in Sweden, the self-proclaimed homeland of social concern, seem to have been worse by far in the early twentieth century than those of Mexican miners in the eighteenth.[55]

Despite many handicaps, Mexicans on both the highest and the lowest levels labored conspicuously to improve the state of the art of mining. In the late seventeenth century the mathematics professor of the University of Mexico, Carlos de Sigüenza y Gongora, became virtually the consulting civil engineer to all Mexico, gaining such international renown that Louis XIV invited him to become a pensioner at Versailles (the invitation was declined). In the 1750's, José Antonio Alzate, the "Pliny of Mexico," educated himself in an engineering curriculum of his own devising and labored earnestly to reform the mines' hoisting

[51] Bernstein, *The Mexican Mining Industry*, 9–10; Motten, *Mexican Silver*, 18–19; Rickard, *Journeys of Observation*, 101–106.

[52] Bernstein, *The Mexican Mining Industry*, 13; Humboldt, *Political Essay*, Book IV, 232–33.

[53] See the section "High-grading" in Chapter VI.

[54] L. B. Bailey's *Indian Slave Trade in the Southwest: A Study of Slave-taking and the Traffic in Indian Captives* is one of the most recent volumes which restates the "black legend" without modification or reservation.

[55] Sandström, *Tunnels*, 320–24.

and pumping methods. About the same period Juan José Ordóñez de Montalvo attempted to upgrade the effectiveness of the patio process.[56] Unfortunately, as matters stood, there was no institution which could bring such men together, grant them teaching facilities, and enable them to sharpen their minds against each others' and those of their students. They worked and wrote in relative isolation, but the deficiency did not go unremarked.

Down in the headings, the humble *barreteros* were also doing what they could. The Indian miners inherited the great stone-working abilities of both the Spaniards and the Aztecs and were soon making masonry an effective and even an aesthetically pleasing substitute for the mine-support timber which Mexico so conspicuously lacked. In this respect it should be remarked that the narrow, high-arched headings of the rat-hole system happen to be of the form (the arch) which is most strongly self-supporting, another suggestion that the seeming primitivism of the *sistema* was actually based in great part upon a realistic appreciation of the facts of life. Mexicans were skilled millwrights and prospectors. Even Rickard, who was prepared to find little good in their work, grudgingly admitted their excellence in drilling, blacksmithing, and porterage.[57] Motten concluded that the Mexican miner of the eighteenth century could hold his own with the best of Cornwall or Germany.[58] Obviously the chief weakness of the *sistema del rato* was the want of education in the lower and middle ranks, but that weakness was nearly universal. The only mining school in existence was that of Freiburg, and the Anglo-Saxon nations were in no remarkable hurry to remedy their own deficiencies in this matter until the middle of the nineteenth century.

It was an idiosyncrasy of most silver lodes that their values came and went unpredictably in a given ore lead, which would one year display a high-tenor shoot, or streak, and the next year thin down to *borrasca* quartz or even pinch out at a fault. Furthermore, the tangled structure of the various lodes meant that a totally new and unanticipated bonanza might be struck by more or less random exploratory drifting. Indeed most of the bonanza strikes after 1750 were made not in new territory

[56] Motten, *Mexican Silver*, 27–32.

[57] Rickard, *Journeys of Observation*, 104–107.

[58] Motten, *Mexican Silver*, 18–19.

but in the vicinity of old and presumably mined-out districts.[59] The leading example was the great La Valenciana, discovered by Antonio Obregón on the strike of the ancient Veta Madre outcrop at Guanajuato.[60]

Such peculiarities inevitably led to habits of thriftlessness, extravagance, and gambling quite similar to the later flush times of Washoe and gave rise to the proverb, "La minería es una lotería." A rich strike touched off conspicuous squandering, and when one was in *borrasca* it was easy enough to borrow on extortionate terms to continue exploration. With a tempting lottery so close to hand, the Mexican capitalists had little incentive to interest themselves in reported strikes in the Far North.[61] Had the metropolitan mines gone into permanent *borrasca* after 1750, that seeming disaster might have forced an interest in the North and promoted the settlement of Arizona, New Mexico, and Colorado. The Mexicans' neglect of the North must be accounted for by geology and human temperament, not by any peculiarity of Hispanic culture.

Such new strikes as La Valenciana were the exception rather than the rule. By 1750 production in the Mexican silver mines was declining alarmingly. The lodes were steadily dropping off in tenor, and the costs of deeper mining were, as always, increasing geometrically with depth. Even though great reserves of in-place ore still remained, it was possible to foresee a day in which most of the mines would be abandoned. The urgent need was to reduce operating costs. That could be done only by achieving greater efficiency, and the Mexican solution bears a strikingly modern ring. The operators organized themselves into the Cuerpo de Minería, a "guild" equivalent to a producers' association, which petitioned the crown for a charter that would grant them remarkable privileges. Among other things the guild wished to obtain development credit on easy terms, the loans being advanced from a revolving fund originally capitalized from the royal *quinto*. It requested termination of the royal monopolies on mercury and salt, a request that was ultimately granted. The guild asked sanction for a general board or tribunal

[59] *Ibid.*, 16.

[60] Rickard, *Journeys of Observation*, 166–67.

[61] For an account of some of these strikes, see the section "Spanish Prospecting in the North" below.

to enact and enforce industry-wide regulation and to put an end to the costly, never-ending litigation which has always characterized precious-metals mining. The guild's most urgent plea, however, was for royal patronage for a college of mines patterned after that at Freiburg, to educate the engineers whose skills were so conspicuously wanting and were so obviously the key to the recovery of the industry.[62]

The Colegio Real de Mínera was established in 1783 by Charles III. Its organization was in the hands of Fausto de Elhuyar, director general of the *tribuna* of the Mining Guild, who sketched out the curriculum, recruited the faculty, and projected an eventual enrollment of one hundred students divided equally among the four years of instruction. The School of Mines was formally opened on January 1, 1792, with eight students and three professors, who taught mathematics, drawing, and French, respectively. Enrollment increased rapidly, but, as present-day mining-school deans may appreciate, remedial Spanish grammar and arithmetic had to be added to the first-year offering to overcome the educational deficiencies of the matriculants. The school gained momentum, adding faculty, students, and courses well on the schedule projected by Elhuyar. Second-year courses included algebra, surveying, dynamics, and hydrodynamics. In the third year mineralogy, geology, and the elements of mining were offered. In the fourth year chemistry and metallurgy were taught by the German professor Louis Lindner.[63]

The swift and sturdy growth of the Colegio Real appeared to herald a new era for Hispanic America, giving it a head start of a quarter-century upon the United States. Sad to relate, however brilliant it may have seemed, the dawn was a false one. In the year the college was opened, revolutionary terror began in France, precipitating world war in which Spain and its dependencies were almost continuously involved. Peninsular Spain was gutted by the armies of Napoleon and by savage guerrilla warfare to the point where its power to control its overseas dominions lapsed. In 1810, Father Miguel Hidalgo y Costilla proclaimed Mexican independence, and the struggle in Mexico became overt. The Mexican Revolution and the tumults of the Reform led to the disruption of mining, of commerce, and of learning. Although the college continued to function for a time, its graduates entered the armies

[62] Howe, *The Mining Guild of New Spain,* 38–51 and ff.

[63] *Ibid.*, 330–42.

of revolution. Its faculty was cut off from the scientific revolution beginning in Europe. In 1867 it virtually ceased to exist.[64] Mining continued after a fashion, but the fashion was the *sistema del rato,* now two centuries behind the times conducted in mines sealed off from the world by stone fortifications.[65]

In sum, it may be said that the Mexican silver industry encountered three major crises: the sulphuret problem of 1545, the depth and low-tenor problems of 1750, and the political crisis of 1810–21. The first crisis was technical in nature and was brilliantly overcome by appropriate technical innovations of Bartolomé de Medina. The second crisis demanded an entirely new order of operations based on science rather than empiricism and appeared well on the way to solution by appropriate educational measures. The third crisis was political, and for it there was to be no solution until the iron dictatorship of Porfirio Díaz brought internal peace and external integrity of a sort in 1877. What the self-appointed critics of the *sistema del rato* failed to appreciate was that in its time it was as effective a technology as the land and the worldwide state of the art permitted. When it became obsolete, the fact was quickly recognized, and well-calculated measures for reform were set in train. That this reform failed to bear fruit was the fault of the political climate of Mexico, and no technical reform could be successful until that climate had been radically altered.

Smelting

When ores were of high tenor and rich in lead—that is, "self-fluxing"—the Spaniard might have recourse to direct smelting, a procedure entirely different from either the patio process or amalgamation. Smelting, too, originated in prehistory, undoubtedly as the product of campfire accident, and by 1500 B.C. had been improved to the point where high-grade surficial copper ores or galena were routinely reduced in crude smelting furnaces. Such a furnace was little more than a simple stack laid up of dry masonry or excavated in a hillside, with air-inlet orifices generally oriented to gain advantage of the draft produced by the prevailing winds. Roughly crushed ore and charcoal were dumped down the stack in alternating layers and fired. Presently a trickle of molten

[64] Motten, *Mexican Silver,* 64–66.
[65] Rickard, *Journeys of Observation,* 113–14.

metal would appear from the lowermost orifice, to be caught and cast in a clay mold. Molds for copper were square with protuberances at each of the four corners. For a long time it was thought that the shape of the ingot was analogous to that of an oxhide and that one ingot equaled one ox in purchasing power, but this theory seems to be unfounded; the four protuberances were merely handles for convenience in porterage. By medieval times the smelting furnace had waterwheel-powered bellows for hotter draft, thus making it a blast furnace, and arrangements for recharging from above to permit continuous runs. It was this system which the Spaniards brought to the New World for nonferrous-base-metal smelting.

When dealing with precious-metal ores, one wants as little contamination as possible, for which reason the Spaniard preferred to use his "Castilian," or reverberatory, furnace, in which the fire is kept separate from the ore charge contained in a *vaso*, or crucible, and its heat is reflected down upon and around the *vaso* by the shape of the walls and confinement of the forced draft. The principle of smelting revolves about some facts which were much refined by empirical experiment but which are easier to explain from the basis provided by modern chemical theory. To start with, nonferrous-mineral values are most often found as a salt, a chemical marriage of the metal atom with a nonmetal atom, which is most often sulphur. These sulphide salts require great amounts of energy to dissociate, or reduce, and the temperature of this dissociation is usually very much higher than the melting points of either of the two components. An alloy, on the other hand, is a chemical partnership of two different metals and, peculiarly enough, usually has a melting point appreciably *lower* than that of either of its two components. It now becomes evident that if the sulphur can be induced to take itself off in some energy-releasing, gas-producing reaction (say, with oxygen) and at the same time its metallic partner can be persuaded to alloy itself with a second metal (preferably one with the lowest feasible melting point) the whole trick can be turned with great economy of fuel and hence cost. This is of some moment where the only fuel available is expensive and relatively cool-burning charcoal, as was the case in the Mediterranean and Hispanic American world.

Smelting of silver sulphurets might be accomplished given three conditions: high-grade ore, availability of fuel, and the inclusion in the ore of an appreciable amount of lead (usually in the form of galena, which, as has been noted, has a high geological probability of occurrence). To this charge in the *vaso* would be added salt and soda. When well heated, the components of the charge would begin to exchange chemical partners. Sulphur combined with available oxygen from the soda and the atmosphere to yield energy and then to depart out the stack as gaseous sulphur dioxide. Silicious gangue rock combined with iron and other rubbish to form slags. The silver and lead formed an alloy whose melting point was even lower than that of lead. At this point the *vaso* was tapped, and its liquid contents poured out into the forehearth, a broad, shallow stone basin. Since the melting point of the slags was far higher than that of the alloy, the slags at once solidified, floated to the top as solid clinkers, and were easily skimmed away. When no more slag appeared, the remaining contents of the forehearth were run off into molds, where they hardened as matte ingots of silver lead. Properly managed, direct smelting was accomplished "as easy as Genesis" (to quote Sam Clemens), unless for any one of an infinity of technical reasons an exception was encountered.

The Spaniard did not fret himself into a nervous breakdown about the refractory ores which did not lend themselves to this process; if the good God did not desire a given ore to be smelted, then it would not be smelted. Therefore, he experimentally tossed concentrates, salt, and soda into his *vaso*, lit his fire, and awaited developments. If he was fortunate, he was presently rewarded. If not, he installed a patio or abandoned the operation. In a later day the Americano smelter operator preferred to give the Almighty a little assistance by paying premium prices for silver-lead ore or concentrates with which to flux his power-blown, water-cooled, coke-fired furnace and to adjust his charge with a Byzantine complexity of pyrites, dolomite, and other chemical equivalents of eye of newt.

Mercury

The freight wagons which rolled across the plains and through the mountain passes to the mining camps of the West carried a thousand items, ranging from carboys of acid for assay houses to French mirrors for houses of another sort. Much more than the proverbial beef, beans, and whisky was needed to support a mining camp, and looming large on the freighters' waybills were flasks of mercury—hundreds of cylindrical iron bottles somewhat resembling modern artillery shells, each containing a trifle more than 76 pounds net of metallic quicksilver consigned to the placer diggings and the quartz mills.[66] The net content was traditional; there is even some reason to believe that it was a rough approximation of the Phoenician talent. For thus far back in history had mercury been a staple of commerce and for most of that time an indispensable adjunct to man's recovery and use of silver and gold.

Mercury, the only metallic element that is liquid at normal temperatures, possesses a surface tension so great as to create the beadlike globules which are its free physical shape. Its eccentricities of behavior attracted the attention of the medieval alchemists, and it is suggestive that they named it after the Roman deity of expedients, evasiveness, and pandering.

By far the most common ore of mercury is cinnabar, whose great weight, crumbly texture, and striking vermilion shade make it unmistakable and interest-arousing. Cinnabar is a simple sulphide and seems to occur primarily in the vicinity of inactive thermal springs, whose moderately hot water leached the mineral from nearby igneous dikes. This water was probably heated in the recent but relatively shallow earth faults which characterize earthquake country and then percolated upward to redeposit its cinnabar content in a diffuse manner through the shale and serpentine or other greatly altered and porous country rock. Cinnabar is seldom found in conjunction with any other metal, presumably because plutonic hydrothermals hot enough to deposit other metals would be so hot as to dissociate, vaporize, and disperse their mercury values into the atmosphere.[67]

Cinnabar has almost certainly been known since prehistoric times. Primitive man was always on the alert for colorful pigments, and vermilion would appeal to the artist and the cosmetician. It could be applied either as a fine-ground powder or as a paste when mixed with

[66] The figure 76 pounds is approximate and variable. The Spanish unit was 76.50 pounds (Thomas Egleston, *The Metallurgy of Silver, Gold, and Mercury in the United States*, II, 807; hereafter cited as *Metallurgy*). The American unit was 76.05 pounds (Ralph S. Tarr, *Economic Geology of the United States*, 255).

[67] Charles M. Riley, *Our Mineral Resources*, 184.

oil or fat. It was still in use as a cosmetic in imperial Roman times, when it was known as *minium*. The classical writers were not too familiar with metallic mercury; the first reference to it was a mention of "liquid silver" in the works of Aristotle. But *minium* was better known. Pliny remarked that it was produced in Spain and noted that even then it was being heavily adulterated with red lead.[68] In modern times mercury was exported to China for the manufacture of vermilion. North American Plains Indians bought large amounts of Chinese-produced vermilion from the fur traders and were particularly fond of applying it to the parting of the hair.[69] The Chinese also used so much lead in their product that the word *minium* came to be used to mean a compound of vermilion and red lead.[70] Unfortunately, both mercury and lead are virulently poisonous (as are most of the heavy metals), and it was probably for this reason that they were replaced during early medieval times by henna, a vegetable dye.

It was early discovered that moderate roasting would reduce the ore and evolve the vapor of metallic mercury, which could then be condensed under water and collected in bulk. The traditional mode of production involved crushing the ore, packing it in ceramic or metallic containers which were luted, or sealed with clay, and kindling a wood fire beneath the containers. By one means or another the quick was then condensed and collected, usually by means of a distillation apparatus similar to those used for the production of more genial fluids (Fig. 21). Storage and transportation of the liquid was accomplished in vessels of glazed pottery, iron, tin, or lead, since it tended to attack and leak from containers of copper or its alloys. In modern times iron flasks have been the customary containers, and mercury has usually been traded and priced by the flask.

There is no record of the first discovery that mercury would amalgamate with silver and gold, although that also was doubtless quite an early discovery (Fig. 22). Mercury combines with gold and silver by a unique wetting process to form a pasty, silvery pseudoalloy of striking—

[68] Aristotle, *Meteorologica*, in *Works* (ed. by W. D. Ross), Vol. IV, 8, 11; Pliny, *Natural History* (trans. by H. Rackham, W. H. S. Jones, and D. E. Eichholz), Book XXXIII, 40.

[69] George Catlin, *North American Indians*, I, 51, 95.

[70] Aitchison, *A History of Metals*, I, 215–16; Bauer, *De re metallica*, 215 note 350.

Fig. 21. Small iron amalgam retort and condenser. From Thomas Egleston, *The Metallurgy of Silver, Gold, and Mercury in the United States* (1887–90).

and, to some, mysterious—properties.[71] When squeezed, the amalgam produces loud, mouselike squeaks impossible to describe but familiar to anyone who has had a tooth filled with this favorite material of dentists. Also, when a cloth or wet chamois bag filled with the amalgam is squeezed, droplets of pure mercury are pushed through the pores, the gold and silver being mysteriously retained within. Separation of the components is done by retorting, carried on with amalgam in much the same manner as the reduction of cinnabar, the mercury being driven off as a vapor while the gold and silver remain behind as retort sponge.

The chief cinnabar deposit known to the ancients was at Almaden, Spain, although the deposit at Idrija, a few miles northeast of Trieste, was found no later than the early Middle Ages. The demand for cinnabar and the working of such deposits was increased by the employment of mercury in goldsmithing, perhaps so early as late Pharaonic times.[72] The process of fire gilding which produces ormolu ware is extremely old, and may even have been independently discovered by

[71] A. M. Gaudin, *Principles of Mineral Dressing*, 477.

[72] Aitchison, *A History of Metals*, I, 215–16.

Fig. 22. Renaissance amalgamation and retorting laboratory. *A*, furnace; *B*, side chambers; *C*, mercury receiver; *D*, retort; *E*, blind head with spout through which water can be poured; *F*, amalgamating concentrates with mortar and pestle; *G*, squeezing mercury from amalgam; *H*, lower part of iron pot; *K*, upper part of iron pot; *L*, leather bag for amalgam squeezing; *M*, melting retort sponge. From Lazarus Ercker, *Treatise on Ores and Assaying* (1951 translation).

the Amerinds of Central America.[73] It is accomplished by carefully brushing a pure-gold amalgam over a copper or silver object so that the amalgam adheres uniformly. The object is then heated circumspectly over a charcoal fire, the heat driving off the mercury to leave the gold as a slightly uneven plating firmly bonded to the base metal. Mercurialism, a disease caused by breathing the poisonous vapors was, of course, the bane of fire gilders, and the process has been prohibited by law within the last century.

The ability of mercury to seize upon and amalgamate gold dust or even to reduce simple silver chloride ores would seem to suggest itself as an ideal placering or milling technique; yet strangely enough there is no record of its use as such until the Renaissance. By the sixteenth century, however, Saxon mining men of the Harz and Erzgebirge mountains were using the quick intelligently in every appropriate stage except inside their concentrators. Yet the fact remains that only about half the metallic content of a placer or free-milling gold-silver ore can be recovered by traditional gravity separation, and this recovery is confined to the larger or "dimensional" flecks of metal. Flood or flour gold will escape in the tailing stream; it is so fine that the water carries it over the gold-saving barriers. Some effort was made to save flood gold by trapping it in fuzzy catch blankets, whose materials ranged from sheep fleece to Brussels carpeting and corduroy, but the utility of the method was somewhat limited. On the other hand, a pool or film of mercury right in the tailing stream would not be much affected physically because of its great cohesiveness and specific gravity, yet nearly all gold particles of whatever size touching its surface would be seized and amalgamated. The amalgam could be removed at leisure with a horn-spoon or spatula and retorted by very elementary procedures, after which the recovered mercury could be used again with little loss.[74]

[73] André Emmerich, *Sweat of the Sun and Tears of the Moon: Gold and Silver in Pre-Columbian Art*, 167–68. Emmerich renders a Scotch verdict on the question of Amerind fire gilding, showing only that mercury has been elicited from certain gilded objects by heating. It must be admitted that the artists could have used native gold amalgams while lacking all knowledge of mercury as such; these amalgams are rare but can be found in placers (Emmons, *Gold Deposits of the World*, 3).

[74] Rickard, *A History of American Mining*, 30–31; [Arizona Bureau of Mines], *Gold Placers and Gold Placering in Arizona*, 99 (hereafter cited as *Gold Placers*).

Despite this one unaccountable blank spot, amalgamation spread rapidly, being introduced to Mexico in 1566 by Fernándes de Velasco as a batch process intended to deal with placer sands, free-milling gold ore, and silver chlorides. Ores were first roasted, dressed, and concentrated and were then introduced to a copper pan whose inner surface was coated with a film of mercury. Salt was added generously, and the wetted mixture was stirred as it was being heated over a furnace. The gold and a little silver were amalgamated by the mercury film which when frosty (fully charged with gold and silver to the natural ratio of one to seven) was rubbed and scraped from the pan.[75] When a fist-sized lump of amalgam had been so accumulated, it was squeezed free of excess mercury and retorted. In small-scale operations in American times, the operator merely placed the lump of amalgam in the hollow scooped from the halves of a large Irish potato. The halves were wired together, and the potato was then baked in a campfire. When cooked, it was opened, and the lump of retort sponge was reclaimed from the hollow. The potato pulp was then manually squeezed to reclaim the mercury, though, needless to say, no one dared make a meal of the residuum.[76]

Pan amalgamation was of little use in reducing sulphurets, but about 1650 some aspects of its operation were combined with the patio process to produce the hybrid *cazo* reduction of refractory silver ores. After being dressed, the silver sulphurets were stamped in the *maza* to the fineness of dusts. These dusts were mixed with water and patio reagents and boiled for four hours in a *cazo*, or kettle. Then they were washed free of gangue, and the cleanup amalgam was retorted. The process was not notably effective, although it would spread to Colorado and the Comstock Lode in the early 1860's simply because no better process was known.[77] Actually, its inventor, Don Francisco Xavier de Gamboa, had missed the boat by very little. Had he bethought himself to roast his dusts with salt and a flux, converting the sulphides to chlorides, and *then* treat them with patio reagents, his recovery percentages would have benefited tremendously.[78]

[75] Aitchison, *A History of Metals*, II, 372.
[76] [Arizona Bureau of Mines], *Gold Placers*, 114.
[77] Lord, *Comstock Mining and Miners*, 81–82.
[78] Gaudin, *Principles of Mineral Dressing*, 477.

The preparation of copper amalgamation plates and pans proved to be a fairly complex art in itself. Charged sheets and pans were carefully rubbed free of the frosty amalgam and then painstakingly washed, freed of stain with mild acid, and washed again. The surface of the plates had to be kept free of obvious scratches, but a matte or dull finish was preferable to any degree of polish. Failure to observe these requirements meant that the mercury film would not adhere properly and under the impact of the tailing stream would begin to slough off. The mercury itself was carefully reapplied in a film which was neither so thick that it would ripple and slough nor so thin that it would fail to catch its allotted quota of gold.

Placer men and millmen also learned that merely bringing any sort of gold dust into contact with casually treated mercury did not necessarily guarantee amalgamation. For that matter, many common conditions badly inhibited the process and demanded appropriate countermeasures. Gold tellurides, or black gold, found in many ores will not amalgamate at all; they must be gravity-concentrated and then roasted to drive off the tellurium. Rusty gold particles, filmed with iron silicates or manganese, are also resistant to the blandishments of mercury. A film of oxidation "sickened" the mercury, but might be alleviated by running acid or coarse salt through the machinery. Perhaps the worst of all the enemies of the process were oil and grease. The smallest trace of oil would cause flood gold to float on the water surface; grease or oil adhered to the crystalline surfaces of metals or metallic salts and prevented the intimate contact essential to amalgamation. Scrupulous cleanliness was the answer; the experienced *gambusino* never ate from his prospecting pan as neophytes might, although a fast field treatment for a pan suspected to be greasy was to heat it to redness over a campfire and thereby burn off the coating.[79]

Spanish Prospecting in the North

On the heels of the Conquest a host of adventurers fanned out across the Americas to discover for themselves treasure-troves in the manner of Cortes and Pizarro. A number of these conquistadors pressed north from Mexico into the future United States as early as 1539–40, seeking

[79] [Arizona Bureau of Mines], *Gold Placers*, 89; Ercker, *Treatise on Ores and Assaying*, 111.

cities of gold under the leadership first of Fray Marcos de Niza, then of Francisco Vásquez de Coronado. Finding that the Indians of the northern deserts and mountains had no such cities, they departed, and so pessimistic were their reports that the North remained unexplored for a generation. Then, in 1582, Antonio de Espejo, hearing a rumor of silver in the Northwest, traveled for that and other reasons up the Río Grande. At its headwaters he turned west, crossed the Sawatch Rockies to the Colorado Plateau, and near the great landmark of the San Francisco Peaks turned back south. He passed down over the Mogollon Rim of the plateau, crossed the Verde River at the base of the escarpment, and in the Black Mountains farther south found mineralized outcroppings which seemed to feature silver.[80] Espejo appears, however, to have reflected upon the want of food and fuel, the distance from Mexico, and the notorious ferocity of the Apache Indians thereabouts. He departed and did not return.

For political reasons as well as because of reports of placer gold in its sands, the Río Grande Valley was colonized north of El Paso by Juan de Oñate, commencing in February, 1598. Despite many difficulties the settlement of Santa Fe took root, but the placer prospects largely failed to materialize. Santa Fe was the gateway to the Back of Beyond, where began the Old Spanish Trail (partly pioneered by Espejo), an enormous-

[80] It is probable that Espejo stumbled upon the great copper lode which would one day become the United Verde and U. V. X. (United Verde Extension, or "Little Daisy") mines and site of the now ghostly town of Jerome, Arizona. Like others, Espejo was deceived by the fact that ground water tends to dissolve metallic sulphides selectively, carry them down the lode, and redeposit them in the rich but shallow concentrations which mark the oxidized zone. Copper is the metal most readily carried down, silver the next, and gold the last. As erosion lowers the mean surface, the gold is exposed first and, unless placered off, will tend to move laterally into the stream beds as dust and nuggets. It is next the turn of the silver to be exposed, gradually reduced to native metal, and then removed by erosion. Copper comes next, leaving iron, and so on. Such a deposit is a layer cake of ore, beginning at depth with a low-tenor copper-sulphide primary deposit, capped by a high-tenor oxidized-copper zone, which is in turn overlaid by a silver-chloride and perhaps a gold-nugget layer. Thus what starts out as a gold placer becomes a silver mine. The silver mine gives way to rich calcocite, "black as coal," and then to a copper-porphyry or similar low-tenor lode. In the U. V. X. portion of the United Verde, the precious metals were gone and the copper so thoroughly carried down that there was virtually no surficial indications of the values below save an outcropping lightly stained with residual iron. Marcus Daley at Butte found the same sequence and capitalized upon it.

ly roundabout trace from Chihuahua to the Pacific Ocean. To improve on this route, in 1776 a party of exploration under Fray Francisco Silvestre Vélez Escalante explored the Sawatch Mountains and the area of the San Juan River. Escalante observed mineral indications up obscure canyons piled with chips of schist and let it go at that.[81] Someone must have heard of this, for American prospectors coming into the Colorado-New Mexico region in 1868 are said to have reported small parties of Mexicans engaged in silver mining and gold placering about Durango, Pinos Altos, and Gila City.

The Jesuits, led by Fray Eusebio Francisco Kino, went to the lower Sonoran Desert in the 1690's, a century after Espejo went out that way. Their reports mentioned mineralization, but their attention was upon matters not of this world. Hardly was their province of Pimería Alta running smoothly when in 1736 occurred the first real mineral rush to the North. It was occasioned by the news that very substantial plaques, or lumps, of native silver were to be found lying about on the ground near the tiny pueblo of Arizonac, just southwest of the present Nogales. Exceedingly rough elements were attracted to this Bolas, or Planchas de Plata, rush, not the least of which was the Spanish crown itself. Learning of the Arizonac, or "Arizona," strike, the crown declined to classify the workings as mines, subject to assessment of the *quinto,* but instead declared the silver a "novelty" and proposed on that account to take all. These first Arizonians did not appreciate the sophistry; they skimmed off the silver and then softly and silently vanished.[82]

About 1750, mission explorers searching for a better and more direct route to the Pacific than the Old Spanish Trail came upon a supergene silver deposit (silver chlorides upon the surface of a porphyritic copper stock) at Ajo, near the trace of the fearsome Camino Diablo, a nearly waterless trail which ran from Jesuit country to the head of the Gulf of Lower California. And in the vicinity of the leading Jesuit mission, San Xavier del Bac of Tucson, two separate silver lodes were known to exist, one on each side of the Santa Rita Mountains. Nothing more seems to have come of these than of the discoveries of Espejo and Escalante, and for the same reasons: the distance from Mexico, the want of every necessity of life, and the danger from the Apaches.

[81] Hubert Howe Bancroft, *History of Nevada, Colorado, and Wyoming,* 339–43.
[82] Rickard, *A History of American Mining,* 258–59.

Far bigger rushes have started on far worse indications than those, but it would seem that the frequent discovery of very rich new ore shoots in the mines of metropolitan Mexico did more than anything else to discourage investment in distant mineral speculations.[83] Just as discouraging, perhaps, was the fact that the administrative and political systems of Mexico were almost perfectly designed to loot a successful prospector who lacked influence in exalted quarters. Though there unquestionably was Mexican mining and even more prospecting in the North, it was conducted in semisecrecy by small entrepreneurs who led laborers into the region, cleaned out obvious surface deposits, and then slipped unobtrusively away to enjoy their modest gains.

[83] Motten, *Mexican Silver*, 16.

CHAPTER IV

California Placering

IN THE MIDDLE 1800's newly annexed California witnessed the first gold rush of the western frontier. The background of the discovery needs but cursory attention. John Augustus Sutter, a Swiss, went to California in 1839, when California was still Mexican territory, to carve out for himself a private agricultural empire along the Sacramento River. He employed wandering James W. Marshall as a general handyman, with a special commission to cooper and wright an undershot water sawmill forty miles up the American River from Sutter's fort, New Helvetia. With helpers recruited from the demobilized Mormon Battalion, Marshall dug the races, erected the wheel, and on January 24, 1848, turned on the water for a brief test. All went well except for some unanticipated earth wash at the head of the tailrace, which threatened to undermine the mill building.

Marshall turned off the water and went to investigate. In the rough ground sluice formed by the race Marshall unwittingly re-enacted prehistoric man's first encounter with gold: his eye was caught by a metallic gleam, and his hand went out for it. He sent an Indian boy for a tin plate, on which he collected about an ounce of lenticular dust and small cubical crystals. Marshall appears to have had no doubt about what he found. That evening one of his crew, Henry W. Bigler, noted in his diary, "This day some kind of mettle was found in the tail of the race that looks like goald."[1]

It is said that Marshall next requested the camp cook, Jennie Wimmer, to boil the flakes in the lye of her soap kettle and that by the next morning the grains showed no discoloration. It is also said that Marshall requested the blacksmith to beat some specks out on his anvil to test

[1] Erwin G. Gudde, *Bigler's Chronicle of the West: The Conquest of California, Discovery of Gold, and Mormon Settlement as Reflected in Henry William Bigler's Diaries*, 88 note 2.

their malleability.[2] Marshall's conviction that the stuff was gold deepened, and on January 27 the mill boss rode with his find to New Helvetia. There he and Sutter closeted themselves for half a day, applying somewhat more effective tests after reading up on the subject in Sutter's *Encyclopedia Americana.* Sutter tried samples with aqua fortis (nitric acid), obtaining the correct reaction—which is no reaction. He and Marshall then used Archimedes' test for relative specific gravity. They put silver coins in one pan of a balance, then added dust to the other pan until the balance hung level. Next the pans were carefully immersed in water. To Marshall's delight the pan containing the coins rose, demonstrating that the dust had an appreciably greater specific gravity. There could now be no doubt—chemistry and physics had united to proclaim that Marshall's find was gold.[3]

The Forty-niners

Sutter (correctly) feared the consequences of the discovery to his growing agricultural empire and attempted to suppress the news, but matters were already out of hand. Virtually every adult male resident of California dropped what he was doing and headed north to the diggings along the tributaries of the Sacramento. The military commander of the California occupation forces, Colonel Richard Barnes Mason, reported the discovery through channels to the War Department, whence it was passed along to President James K. Polk. Polk, who was desperate for some good news to offset party quarrels and the sour national reaction to the unpopular Mexican War, gave Mason's optimistic report more than adequate publicity. The public had the winter of 1848–49 to digest the news, and by the spring of 1849, fifty thousand Americans had begun crowding the sea routes and overland trails to California, every man of them intent upon the gold.[4]

How had the gold got there? The forty-niners eventually discovered that it had derived from a variety of sources. Some came from statisti-

[2] Rickard, *A History of American Mining,* 23–24.

[3] *Ibid.,* 88–91; Daniel Tyler, *A Concise History of the Mormon Battalion in the Mexican War, 1846–1847,* 333–34 (hereafter cited as *A Concise History of the Mormon Battalion*).

[4] Rodman W. Paul, *California Gold: The Beginning of Mining in the Far West,* 24 (hereafter cited as *California Gold*).

cally infinite dispersion among the billions of tons of granitic rock which made up the core of the Sierra Nevada.[5] A hundred tons of such rock might contain only a half thimbleful of tiny golden specks scattered at random within it, but the mountain heights and the rainy climate had acted as a natural concentrator, milling and freeing the gold, then washing it down to the placering grounds after some interesting delays en route. More of the gold came from quartz lodes which had intrusively followed faults and fissures created as the granitic core of the Sierras rose and tilted to the west. The major lead was the "Mother Lode"[6] which parallels the San Joaquin River, being traceable as white outcroppings along the mountain flanks or foothill crests for sixty miles from the Cosumnes River to the Mariposa River. This lead, the "first among equals," was a dike some yards wide, penned between the hanging granitic wall rock of the Sierras on the east and the foot wall of the uptilted metamorphic shales and slates of the erstwhile sedimentary formations on the west. As is so often the case, the main quartz dike itself did not contain much of value, but its side branches—"shoots"—and the contact zone between it and the foot wall contained valuable mineralization. This habit is peculiar to intrusive acidic dikes.[7] Where such leads can be observed from the air, as in desert country, it is obvious to the naked eye that the mine dumps are found on either side of the lead but are seldom directly upon it (Fig. 23).

Like the country rock, the outcroppings of the quartz leads were naturally milled and concentrated, their gold content being washed down to a series of archaic river beds. The "Big Blue Lead" of the hydraulickers, which supposedly ran from north to south, was roughly parallel to but appreciably east of the present Sacramento River.[8] The Sierras tilted farther west and rose in elevation, lifting and draining the

[5] Granitic rock contains many metals but is not worth milling because of the infinitesimal per-ton yield. Another example of natural concentration is the brown iron "paint" of desert-weathered granite boulders which, when cracked, show the natural light gray within.

[6] The "Mother Lode" is subject to a variety of definitions. The quartz dike mentioned parallels the San Joaquin River, but some define it as the "Mother Lode country" of the placer-camp belt east of the Sacramento River. Cf. Paul, *California Gold*, 40–43.

[7] Hubert Howe Bancroft, *History of California*, VI, 383.

[8] Lakes, *Prospecting*, 120. The "Big Blue Lead" was a geological fiction supposedly traceable from Alaska to Mexico by its pyrite-colored lower gravels, and

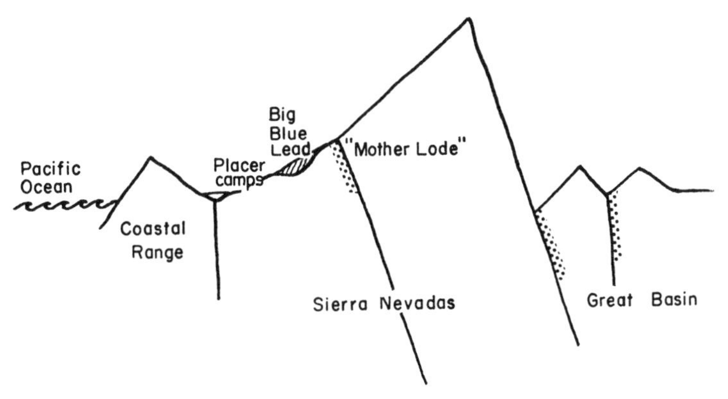

Fig. 23. Schematic diagram of the California–Great Basin profile, looking north (not to scale).

river beds until they became no more than banks of gravel resting high and dry on the flanks of the ranges. The mountain streams continued to flow west, but they now cut across the belt of detritus and passed on down to the new Sacramento River valley, picking up and redepositing some of the placer gold the ancient rivers had contained.

In Recent (Quaternary) times, basaltic volcanic activity had covered this gravel bank in such places as Table Mountain on the Stanislaus River with one or more sheets of lava, but the bank remained substantially in place.[9] Therefore, what the forty-niner was doing with pick and pan along the streams—or would do with hydraulic monitor up along the Big Blue Lead—was merely the final act of a geological drama which had already been ages in preparation. The free gold, or gold tellurides [10] were liberated by natural erosive processes: the driving

was considered by the Californians to have been of one body. For the true but complex genesis of the auriferous gravels of the Oroville, Folsom, and Yuba districts, see D'Arcy Weatherbe, *Dredging for Gold in California*, 9–15.

[9] Bancroft, *History of California*, IV, 373–74. The highly liquid basaltic flows tend to overrun the country in "tongues," which harden and whose exposed edges show the typically crystalline form of basalt. Later erosion removes the land except where the very durable basalt layers shelter it as with an umbrella. The consequence is a table or mesa, characteristically flat-topped, standing well above the mean surface and prominently displaying its basalt capping.

[10] Tellurium, despite its metallic-sounding suffix, is a semimetal of the oxygen-sulphur-selenium group and, peculiarly enough, will form an insoluble salt with

Sierra rains, the riving frosts, and the expanding tree roots, which had all gone about their business of splitting cliffs into talus slopes and wedging boulders away from bedrock. As each block fell, gravity and impact broke it up into smaller fragments. These fragments in turn fell into the extraordinarily effective wet grinders of the mountain streams,[11] which by mutual impact and abrasion reduced them into sands and silts as thoroughly as could any man-made mill.

At this point the gold was freed from the surrounding gangue, but its ductility preserved it from being in turn atomized. When crushed between two stream-bed cobbles, the gold flattened into scales, or even melded together, but did not break up. Being chemically inert, it was unaffected by the natural acids in the stream water, which would otherwise have dissolved it and washed it away in solution. Even so, as much as half of the gold occurred in grains so tiny that the water's turbulence and surface tension buoyed them up and carried them to the sea. The placer man thus distinguished between heavy gold and flood gold; the former was larger and found next to the bedrock, whereas the latter, being fine and almost impalpable, was renewed in supply by the seasonal runoff but was scarcely profitable to recover.

What remained was carried along by the streams, intermingled with sands and silts, but clearly distinguishable as gold. In one year a stream might sweep down only a handful of such dust, but what is one year when the mountains have millions in which to do their appointed work? The grains of gold advance during the time of the winter rains, halt during the summer dry season, then once more resume their erratic journey when the next freshets come. In a few centuries or millenia the individual particle of gold freed halfway up a mountainside might either make its way to the pediment below or else be lost beyond re-

gold, which sulphur itself will not. Tellurides, chlorides, and cyanides are virtually the only common salts of gold; tellurides persist because of their insolubility, whereas chlorides and cyanides are remarkably soluble, and so are never found in the natural state. Gold tellurides, or "black gold," can be recovered by water placering, but will not amalgamate. The usual method of reducing tellurides is by roasting or smelting. The stigma of the telluride smelter or miner is the remarkably offensive "tellurium breath," which comes from inhaling the element.

[11] Mountain streams are swift, and in this connection it is significant to note that the energy of water available for erosion or abrasion is directly proportional to the square of its velocity. Thus a stream flowing at eight miles an hour can do sixty-four times the erosive work of a stream flowing at one mile an hour.

demption—wedged between cobbles, fallen into a crevice, or stranded when the stream changed its course.

The process approaches its climax as the stream leaves the steep mountain grades and perceptibly slows in the rolling pediment. The water abruptly loses most of its energy and is no longer able to command the destiny of the high specific gravity of the larger pellets of gold. The pellets sink to the rocky stream bed in company with the larger cobbles of rock. The current rolls and tumbles them onward for a time, but sooner or later they settle permanently. Finer particles go farther but then in turn sink as the stream slows more and loses yet more energy. The rule is nearly exact: the closer to the mountains one goes, the relatively greater the size of the gold dust he discovers. Down near the stream's mouth, as where Marshall made the discovery at Sutter's mill, the dust is fine and widely dispersed.

There is a final step. After they have been stirred vigorously, solid objects tend to sort and classify themselves neatly in accordance to weight and size. The particles of higher specific gravity sink; the larger and lighter objects rise. As seasonal freshets disturb and roil the stream bed, the coarse pebbles and shingle rise to form a cobbled stream floor; the finer sands, including the gold dust, sink. Gold, being heaviest of all, descends toward the bedrock and there reposes in a layer invisible to casual inspection but far more concentrated than at any other stratum of the alluvial detritus. This layer is the pay streak, and it is richer within the transition zone, where the mountain streams slow their rush, than ever it was in the mountains themselves, or will be down in the alluvial plains. Historically, this was the belt of the gold diggings and the placer camps. There with pick and pan a lucky man might in a week take out a fortune in gold dust, while his neighbor a few yards distant on the same stream might encounter only the paltriest show of color for his pains. The sole difference between them has been incalculable, statistical chance. Now and then a nugget of considerable size might be found. The first-comers tended to regard this as a distinct chunk broken away from the "Mother Lode," which they thought existed like a great yellow curbstone outcrop somewhere high in the Sierras. In fact, a nugget is either the product of the aforesaid pressure melding[12] or else

[12] Pressure melding can occur even in stacked and rolled gold coins which have been overlong in storage.

the deposit in some vein cavity of gold brought thither, atom by atom, in consequence of slow chemical seepage, which was then liberated as a unit—as its irregular, rounded surface, devoid of cleavage planes, proclaimed.

Placering

Sutter's private barony was overrun by gold seekers, who turned its stream banks into something resembling the mountains of the moon. By the end of 1849 the placer camps stretched from Mormon Bar, at the headwaters of the Mariposa River, two hundred miles north to Rich Bar, on the north fork of the Feather River. Though in the first flush a good many forty-niners merely crevice-hunted with knives or searched aimlessly about for obvious nuggets, most of them appear to have solicited the advice of the Mexicans and Chileans, who wasted no time in putting in an appearance. A few days' experience indicated that one had to dig for the pay streak rather than poke promiscuously about and that there were places to dig and places at which to refrain from digging. A straight reach of stream, for instance, was to be avoided in favor of places where the current slowed. Likely spots would be those where the grade decreased or where the channel widened, deepened, or joined another stream. One of the best sites of all was the upper end of a sandbar building out at the inside bank of a bend or in the downstream mouth of an oxbow left by a change in course of the creek. Shallow gravel bars were easier to work than were deep banks—a useful indication of the former were bedrock ridges protruding through the water or the bank itself. Since transverse rock ridges or potholes were natural riffle bars, they were carefully investigated. Of course, other heavy minerals, such as garnet, black magnetic sand, and hematites, tended to accumulate in the same places; and since their presence was easier to detect by casual inspection than was that of gold, these minerals were regarded as useful signposts of the pay streaks or spot accumulations.

Where water or construction materials were scarce, some placer men used "bed-sheet" dry washing: a shovelful of dry sand was scattered over a sheet or smooth blanket, which was held by miners at each end. At the approach of a breeze the soil was repeatedly tossed up to be air-winnowed. A variation of this procedure was to place a shovelful of sand on the sheet and ladle water on the blanket. The water washed over the sand and carried it away, as with a planilla board. Dry washing was never greatly favored for the obvious reason that it was wasteful and slow; moreover, when one was cleaning up the residues, the physical puffing tended to produce spots before the eyes. The process was never totally abandoned, however. Bartenders and others who accepted gold dust in way of business possessed themselves of a blower, which was a square, funnel-shaped box. Into this box the dust was poured, to trickle forth from the small end into the balance pan. The blower puffed cautiously across the outgoing stream to rid it of the fine silt which was commonly present and did not worry too much about losses, since it was the other man's dust which was being so treated. Besides, the lost dust might later be recovered at leisure from between the floorboards of the establishment. Hence the greater the carelessness the greater the profit.

In California dry washing gave way almost immediately to panning, allegedly introduced at Coloma by Isaac Humphreys, of Georgia.[13] The theory of panning could be grasped instantly by the novice, although its operation was not as simple as it appeared. Indeed, annual national panning championships are still held to reward the greatest degree of speed and finesse in the art. The contestant takes pan in hand, fills it with two inches of sand, and hunkers down ungracefully in a foot or so of water. The pan is shaken, hand-picked of pebbles, and swirled with its top edge barely submerged so that the sand is washed over the top edge, while the operator restrains errant gold colors from departing in the same manner. At the end a few drops of silty water, some black magnetic sand, and a thin tail of golden specks are—presumably—left in the drag of the pan, whence they are taken and added to their predecessors in a tin cup or similar receptacle. At the end of the day this cleanup is dried, puffed free of silt, and poured into the cylindrical buckskin poke or purse.[14] In gold-rush days fifty washed pans were considered a man's daily labor quota.

[13] The pan was the flat-bottomed "American" pan, not the conical Spanish batea, which it supplanted. Many Georgians and Carolinians were adept at placering, having learned the art in the Appalachian South, where placers and soft outcrops had produced about twenty million dollars in gold from 1799 to 1860. See Rickard, *A History of American Mining*, 18–19.

[14] In early days argonauts with richer claims or more highly refined tastes pre-

The first panning utilized closely woven baskets of Indian manufacture, whose price was quickly driven up to fifteen dollars by overwhelming demand. Later, better pans of wood, tin, or iron were introduced, and the pan rapidly became standardized in sheet iron, eighteen inches in diameter and four inches deep, with sides slanting outward at an angle of thirty-seven degrees (such a pan is still a standard item of stock in many western general stores). One Alexander Stephens (who may have been a member of James Marshall's work gang[15]) is said to have nailed rockers beneath a large wooden bowl pan and attached a handle to its side, an improvement which saved much energy because it permitted the user to stand upright and rest the bowl on the ground while using it.

The argonauts got at the placer sands by digging, usually all the way to bedrock. Sandbars and low-lying gravel banks were the obvious and natural places to begin—as was any place where the stream abruptly slowed and demonstrated the fact by dropping a substantial amount of detritus. With pick and shovel the operator scooped out a gopher-hole pit, pausing now and then to wash a test panful to check his progress. When the pay streak was reached, the layer was carefully excavated and extended until it played out or reached the boundary of an adjoining claim. The process, which resembled small-scale open-pit or strip mining, entailed the removal and disposition of an enormous amount of valueless overburden.

To save time and energy, the miner occasionally resorted to "coyoting" —digging one shaft down to the pay streak and then drifting outward into the streak with a series of radial tunnels like the spokes of a wheel. In friable or "heavy" ground this process could be deadly dangerous, and cave-ins undoubtedly accounted for a good many lives (Fig. 24). (Later on in the Yukon, coyoting was used almost exclusively; it was warmer than open-pit placering, and the permafrost supported the ground without need for much timbering, though the frozen ground required fire setting to thaw it for excavation.) Upon exhaustion of

---

ferred to go to the local abattoir to procure for this purpose an appendage of ampler capacity for which the slaughter bulls no longer had any use yet which had been admirably designed by nature for the retention of valuable possessions, whether animal or mineral.

[15] Tyler, *A Concise History of the Mormon Battalion*, 333.

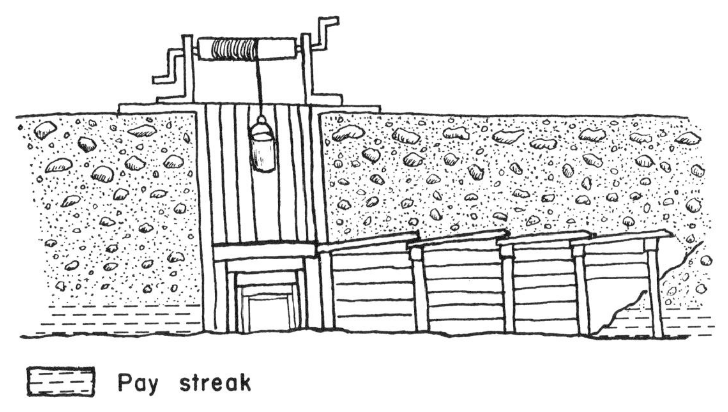

Fig. 24. Coyoting—advancing the heading by forepoling.

the pay streak, the bedrock was cleaned of debris, and any cracks and potholes therein were carefully investigated with a knife point. Luck was everything in the matter; a fortunate man might take out fifteen thousand dollars in a week, but a claim was regarded as breaking even when it produced an ounce of dust (fifteen dollars) a day. Anything less was "Chinaman's diggings," since only a frugal Oriental could make enough to eat on less than an ounce a day, gold-rush prices being what they were.

Mining Claims

Once the rush was under way, a multitude of technical and legal problems almost immediately arose. What constituted a claim, and how did one gain and retain possession of it? How about water supply? What was to be done with excavated spoil? How could one speed up the washing process—assuming that adequate water, materials, and pay dirt were at hand? What of the deposits presumed to lie under the main course of the stream itself? How was one to deal with the high-and-dry gravel banks of the Big Blue Lead? For that matter, the very storage, transmission, and evaluation of gold dust was a major issue. The forty-niners attacked each of these questions with characteristic verve, energy, and realism, and it was hardly more than a year before all of them had been effectively answered.

Both expedience and common law indicated that mineral wealth should *not* be regarded as an integral part of the bundle of rights which constituted normal ownership of real property. Rather, the finder of gold had a perfect right to the gold he found, quite independently of the nominal landowner's equity. He also had right to water and access, and so much the worse for Mexican grantees like the unfortunate Sutter. The common law also held that in a land otherwise devoid of appropriate law or a lawgiving body (as California practically was), the free citizenry might legislate for its own needs and that, as long as this legislation was reasonable and equitable, subsequent formal sovereigns must recognize this prior legislation as valid. Being instructed to this effect by the many lawyers among them (each of whom had virtually memorized Blackstone on Coke), the argonauts proceeded to organize folk moots, or "miners' meetings," in which placer law was debated and ratified by vote of all adult males present. Even though a miners' meeting professed to legislate only for its own district, the claim legislation proved remarkably uniform in spirit throughout the California diggings.

As a rule, a claim was fifty to one hundred feet long and extended along the thread (center) of the channel of the watercourse back a distance upon the bank sufficient for working purposes. A claim had to be located and a location pit put down, its landward corners clearly marked with stakes, and recorded as such with an appointed or elected recorder of the district, who received a fee for his trouble. No man could hold more than one claim at a time in one district, although partnerships might hold as many contiguous claims (termed an association claim) as there were partners. Unless a claim was worked by its holder a specified minimum length of time, the claim was considered abandoned and could be relocated by someone else.[16] No claimant had the right to engross, or take more water than he needed for his immediate purposes, or to dump spoil upon the claim of his neighbors or otherwise impede him. Any disputes could be settled by reference to the recorder, to referees, or (in the case of unusual circumstances) by appeal to the miners' meeting for final adjudication. Willful violation was punishable (again, by majority vote in miners' meetings) by anything from whipping and exile to hanging.

[16] In 1848–49 the accepted sign of active possession of a claim was the presence of mining tools left in the pit; this sign ensured against encroachment for the ten days. Cf. E. G. Waite, "Pioneer Mining in California," *Century Magazine*, Vol. XLII (1891), 134.

Water supply was dealt with in similar fashion. Since all placering depended upon water, it was established as a principle that this commodity should be as evenly divided as possible throughout the stream district. Those who wished to build dams for impoundment or river-bed working were at liberty to do so, provided they made no attempt to monopolize the supply. Infringement was met with instant demolition of the offensive structure; if he was lucky, the would-be monopolist was let off with a stern caution. Spoil disposal likewise was regulated with the view of making a man kill his own snakes, and quite a few popular tribunals sentenced a defendant to remove his spoil from the property or watercourse of his neighbor by the sweat of his brow.[17]

Washing

Dealing with increasing quantities of low-grade pay dirt became pressing almost immediately. Once down to the pay streak, a man with a shovel could remove it faster than two or three men could pan. The solution was the four-man team, armed with a cradle or rocker. In principle, the cradle was an abbreviated riffle box mounted upon transverse rockers. Atop one end was a removable hopper equipped with some sort of mesh screen or perforated iron "riddle" of punched plate. A slanting canvas apron ran from the "low" side of the hopper down to the "high" side of the riffle box. While two men dug, the third dumped the pay sand into the hopper, following it with buckets of water, as the fourth partner vigorously worked the rocker handle to and fro. The riddle scalped off the gravel, while the water and agitation worked the sands down the apron, over the riffle bars, and out the end of the cradle. From time to time the hopper was lifted, and its accumulation of gravel was dumped to one side. At appropriate intervals the canvas apron (which acted as a sort of catch blanket) was washed out, and the riffle bars were cleaned with a horn spoon. It was a crude device, but it worked, and even an unskilled ribbon clerk could cobble a rocker together in short order. It eliminated squatting haunch-deep in icy mountain streams for hours on end, as panning had required. The pan was forthwith retired from every use but prospecting and cleanup (Fig. 25).

The cradle helped washing outpace digging. The next step was to

[17] Charles Howard Shinn, *Mining Camps: A Study in American Frontier Government*, 103–18, 233–38 (hereafter cited as *Mining Camps*).

Fig. 25. The cradle, or rocker. From Thomas Egleston, *The Metallurgy of Silver, Gold, and Mercury in the United States* (1887–90).

begin exploiting the stream beds by river mining, whose usual method was the erection of dams or wing dams to dry and expose the downstream channel sands, together with a large flume (wooden aqueduct) to bypass the normal water flow to one's neighbors, lest they take umbrage. Undershot water wheels, variations of which were called Chinese pumps, could be set in the flume to power endless-chain bailers to keep the working area dry (Fig. 26). Since gold tended to accumulate along the banks and bars, rather than in the channel where the current had been swiftest, river mining generally failed to return more than the investment.[18] Nevertheless for a time it was so popular that it was said that about 1851 the Sacramento River failed to touch so much as a foot

[18] An exception was the experience of one Colonel James, who river-mined quite successfully at a point where upturned edges of slate strata crossed the stream bottom to form a series of deep and profitable natural riffle bars. Gold seldom collected on clean rock bottom, however (Waite, "Pioneer Mining in California," *Century Magazine*, Vol. XLII [1891], 132).

Fig. 26. River mining in the California diggings. From Louis Simonin, *La vie souterrane* (1867).

Fig. 27. River mining. The stream has been dammed and diverted by a bypass flume. The undershot wheels in the flume are driving Chinese chain pumps. From Thomas Egleston, *The Metallurgy of Silver, Gold, and Mercury in the United States* (1887–90).

of its original bed, being then conducted from source to mouth by new ditches or bypass flumes (Fig. 27).

One notable exception to the rule that the pay streak went down as far as it could go was found in the sand-filled potholes of the stream channels revealed by river mining. In these potholes, particularly those of the larger size, the dust was unaccountably found in a high, thin streak just beneath the lip of the hole. A reasonable explanation was that natural whirlpools had carved the potholes and disported themselves therein. It is a law of hydraulics that objects of low specific gravity tend to congregate in the slowly moving center of such a swirl while those of greater mass are thrown out to its periphery by centrifugal force. Since a section of a whirlpool is conical in shape, the zone of greatest centrifugal force corresponds nicely with the high circular streak where the gold was actually found.

The combination of an available head of water, temporarily unlimited sands, and Mexican tutelage produced a widespread revival of the sluice or riffle box, sometimes confused even by the forty-niners with the long tom. In its crudest form the sluice was merely a shallow trench, a ground sluice, whose bottom was riffled with holes, gravel riffle bars, or even cobbles (Fig. 28). A more refined version was the timber sluice, which

Fig. 28. A ground sluice. From Thomas Egleston, *The Metallurgy of Silver, Gold, and Mercury in the United States* (1887–90).

was little more than the old Mexican trough, approximately twenty feet long and eighteen inches deep and reinforced at intervals with top and bottom cleats. The interior was provided with "Hungarian" riffle bars arranged in downward-pointing V's or square U's. The device was built with a slight longitudinal taper so that a number of individual sluices could be interlocked, telescope-fashion, being extended upstream as the diggings were advanced. With a wooden flume to lead in the water and a screened hopper, or tom, to reject gravel, the sluice would process as much sand as it had room along its sides for men to stand and shovel it in.[19]

The long tom could be used individually or at the head end of a sluice line. Its upper component was a broad wooden deck, tapering somewhat to a riddle sheet or mesh screen at the low end. This soon gave way to a sluice, locked in a few inches below the tom's outlet. The tom was developed to wash very coarse pay dirt; its upper deck could hold a great deal of gravel, yet be quickly cleared of it with a few casual sweeps of a scoop shovel (Fig. 29). Toms fell out of use with the invention of the sluice fork, which resembled a hayfork but with more numerous, flattened tines. This device could scalp off the gravel in a

[19] *Ibid.*, 140.

Fig. 29. A long tom. Right to left: flume, tom, riddle plate, sluice. From Thomas Egleston, *The Metallurgy of Silver, Gold, and Mercury in the United States* (1887–90).

sluice quite readily.[20] Whether tom, riffle box, or sluice was used, all had catch blankets at the lower end of the riffle-bar deck; for the Mexican sheepskin the argonaut substituted corduroy or machine-made Brussels carpeting (Fig. 30). Finally, mercury might be poured above the riffle bars when available and appropriate (the presence of iron or manganese-coated dust rendered this measure inappropriate).[21]

The New Almaden Mercury Mine

American placering in California might have remained at the level of that of antique Colchis had it not been for the providential discovery within California itself of the mercury deposits which the forty-niners required for inside amalgamation. About that time the Rothschild banking interests held control of the cinnabar deposits of Almaden, Spain, using their corner to drive up the price of mercury to $114.50 a flask despite competition from Idrija and Huancavelica. That might

[20] Hughes, *Pioneer Years in the Black Hills*, 76.

[21] Schematic diagrams differentiating between tom and sluice are to be found in Arizona Bureau of Mines, *Arizona Gold Placers and Gold Placering*, University of Arizona *Bulletin*, Vol. XXIII, No. 1, 101, fig. 10, 106, fig. 12.

Fig. 30. Renaissance placering equipment. *A*, miners emptying ore into the screen; *B*, the screen; *C*, washer; *D*, upper and lower chutes; *E*, blanket sluices; *F*, workers stirring mud; *G*, barrel in which blanket is washed. From Lazarus Ercker, *Treatise on Ores and Assaying* (1951 translation).

have crippled the American mining frontier, but back in 1824 Secundio Robles, probably a civil functionary of California, had observed the Indians of Santa Clara County using vermilion paint of their own production. He denounced the deposit, which he ultimately named New Almaden. Not surprisingly, it sits astride the notorious San Andreas Fault system, as does the nearby New Idria deposit in San Benito County. The Robles family lost interest in their possession and transferred a considerable share to the Californio soldier-politician Andrés Castillero. Castillero in turn assigned a large percentage of his equity to the British-Mexican firm Barron, Forbes and Company of Tepic, one of whose principal shareholders, James A. Forbes, had been British consul to California.

The gold rush provided such a boom market that Barron, Forbes commenced massive production in 1850, to such good effect that the California market price dropped to $47.83. By report the first roasting furnace employed iron tubes adapted from musket barrels, but in all other respects Barron, Forbes worked the deposit by the *sistema del rato*. Only the low wages paid the Mexican miners and the richness of the ore (in places, it was said, when one drove a pick into a pocket of cinnabar a pound or more of liquid quicksilver would spurt out) enabled the firm to clear a profit.

It has been said that litigation invariably accompanies precious-metals mining, and New Almaden was no exception. There began a decade of suits against Barron, Forbes, instigated by the Quicksilver Mining Company, an American-owned corporation backed by the Bank of California, which sought to establish a monopoly in mercury production. In 1864 the United States Supreme Court rendered an ambiguous verdict, but in the end Barron, Forbes sold out to the Quicksilver Mining Company for a comparatively paltry $1.75 million. As a consolation prize its distributing subsidiary, Barron and Company of San Francisco, was made the sole outlet of what was now, in fact, a total monopoly of the product. In 1874 the Quicksilver Mining Company drove the market price to a high of $105.18 but also modernized New Almaden extensively. In its peak year the mine produced 200,000 pounds of the quick, and the company grossed an average of just under $1.00 million yearly during the fifteen years of its monopoly. Among other innovations the company introduced the "American" mercury flask (actually of British manufacture). Unlike the long-necked Spanish flask and the short-necked Mexican container, it was an almost true cylinder with a flat base and a top end curved only slightly, closed with a threaded and semi-recessed plug. *Borrasca* at New Almaden and a revolt among its participants terminated the Quicksilver Mining Company stranglehold in 1875.[22]

Sluices

A cheap variation of the sluice, possibly used in areas where the gold colors were exceedingly minute, abandoned amalgamation and riffle bars altogether in favor of a plain, wide bottom of tightly stretched canvas. This canvas apron slackened slightly when under load, creating a long, shallow indentation or pocket in its center line from front to rear. Presumably the water flow washed the sands and silts over its lateral edges as well as its lower end, leaving the dust entangled in the weave or deposited in the pocket in a manner analogous to a vanner. Daguerreotypes show sluices and toms with what appears to be this arrangement, although the smooth lower surfaces may be only catch blankets or carpeting. One may conclude that the canvas apron might be a trifle more efficient than riffle bars would be in trapping very fine grained dust or flakes but that adjustments of the water flow and sand feed would have to be made with extreme care in order for it to work at all.

Having solved to their satisfaction the questions of digging and washing, the forty-niners were now up against the problem of obtaining adequate amounts of water, particularly in view of the seasonal nature of the rains,[23] for the working of placers any distance from the streams. It was obviously impracticable for one claimant to erect, say, ten miles of flume in order to bring down water only for himself; therefore, in-

[22] Paul, *California Gold*, 276; Egleston, *Metallurgy*, II, 799; Humboldt, *Political Essay*, Book IV, 283; Johnson, *The New Amalden Quicksilver Mine, passim*; Bancroft, *History of California*, VII, 656 note 20; Arthur C. Todd, *The Cornish Miner in America: The Story of Cousin Jacks and Their Jennies and Their Contribution to the Mining History of the American West, 1830–1914*, 83–103 (hereafter cited as *The Cornish Miner in America*).

[23] The "Express Panic" of 1855, in which Adams & Company, as well as several other California banking houses, went broke, was in part due to a severe drought which prevented placering and so denied them the gold remittances on which they were depending to meet their commitments.

dependent water companies were quickly organized to service whole districts. They built their dams and flumes, surveying for the latter with homemade plane tables, and establishing flume grades with flat- or square-faced whisky bottles nearly full of water to serve as crude but consistent leveling devices. One statistic will serve: the longest of the systems was the 247-mile line of the Eureka Canal in El Dorado County, California. The water companies then charged either flat rates or percentages of the cleanup from all who used their services. For a time rates were set by the "Tom head"—the two-inch stream of water which experience had indicated was the optimum rate of flow through a sluice. Soon was substituted the since-hallowed "miner's inch"—the amount of water which can flow through a hole one square inch in size, said hole being 6½ inches below the upstream water surface.[24] If not specified otherwise, the traditional royalty was 50 per cent of the cleanup from the first user of the water, 40 per cent from the second, and 30 per cent from all subsequent users (Fig. 31).

Cleanup amalgam was bought by weight and estimated gold content by itinerant purchasers who usually gave fourteen to sixteen dollars an ounce, but the average argonaut was quite capable of dispensing the dry and winnowed dust which he collected otherwise. Balances promptly appeared in every store, saloon, and express office in the diggings. A whole body of stories of mild chicanery arose in connection with the custom of dealing in dust. Small change, for instance, was the pinch of gold dust extracted from the buyer's proffered poke by the seller. A pinch was considered the equivalent of one dollar,[25] and discrepancies were supposed to average out in the long run. Since hardly anything in California of the 1850's cost less than a dollar, this system was adequate for quite some time. Sharpers, however, let their thumb and index fingernails grow long and manicured them frequently; one merchant customarily pinched a conical button between thumb and forefinger when not otherwise occupied. It is said that one bartender greased his hair liberally, passed his fingers through it after every sale, and then took a

[24] The time during which the stream flowed was established at ten hours (one working day), but the reader is cautioned that modern definitions of the miner's inch vary considerably from the early standard (Waite, "Pioneer Mining in California," *Century Magazine*, Vol. XLII [1891], 140).

[25] In some camps, twenty-five cents. The difference may be attributed both to custom and to variation in the size and fineness of dusts.

Fig. 31. An early California water-company flume. From C. Le Neve Foster, *A Text-Book of Ore and Stone Mining* (1897).

very profitable shampoo each night after work. Urchins paid for the concession of working saloon and hurdy-gurdy floor cracks with the wetted heads of pins, much as small boys today fish through storm-sewer gratings for small coins with fishing line, weight, and well-masticated chewing gum.

Disposition of Gold Dust

Although any thorough discussion would enter the field of specialized numismatics, it is worth mentioning that certain San Francisco merchants and assay houses soon fell into the logical practice of minting gold dust and retort sponge into slugs for convenient use, exactly as had residents of Ionian cities of Asia Minor three thousand years previously. Intrinsic value was equal to, or slightly exceeded, face value, and a good many of the pieces were given an octagonal outline to distinguish them from "real" money. The slugs attained some local popularity, particularly in the gaming houses, but were quickly either melted down for their gold content or else placed in private hoards by the product of the United States Mint established in the city. It would be logical to suppose that "bearer" paper of the express companies, issued as receipts against gold dust deposited with them, might well have circulated in the same fashion as did tobacco-warehouse receipts in colonial Virginia, but I have not found any source to substantiate this presumption. The bulk of the gold which came down from the diggings was certainly fire-refined quickly and cast into bullion ingots for ocean shipment via the Isthmus of Panama to eastern banks and correspondents.

So great was the eastern demand for gold and the California demand for consumer and capital goods that the bullion output was shipped almost as soon as it had hardened in the mold. Although the gold output was impressive in the absolute sense, amounting to $560 million in the decade 1850–59, it worked out to an annual return of about only $250 an argonaut—a smaller real income (considering prices) than the argonauts would have made had they remained in the States to work for prevailing day wages. Subtract one-fifth of the gross amount to the account of foreign miners' remittances home, and to East Coast interest paid in specie to European capitalists, and the total figure is still $450 million, a highly significant increase in the national currency, just at the time when such institutions as the railroad were beginning to demand really huge amounts of investment capital. In short, the gold rush could not have come at a better time, from the viewpoint of the American heavy-industry revolution.

Plate 1. A *malacate*, or horse whim. From Thomas A. Rickard, *Journeys of Observation* (1907).

Plate 2. An arrastra and its millwork, driven by an overshot water wheel. From Thomas A. Rickard, *Journeys of Observation* (1907).

Plate 3. A Chilean mill in operation in Mexico. From Thomas A. Rickard, *Journeys of Observation* (1907).

Plate 4. A Mexican silver patio. From Thomas A. Rickard, *Journeys of Observation* (1907).

Plate 5. A close-connected bucket-line gold dredge in operation. From D'Arcy Weatherbe, *Dredging for Gold in California* (1907).

Plate 6. Machine man, chuck tenders, and surveyors at work in a Cripple Creek, Colorado, mine stope. Photograph furnished by Arthur H. Lidman.

Plate 7. Cripple Creek miners eating lunch. Photograph furnished by Arthur H. Lidman.

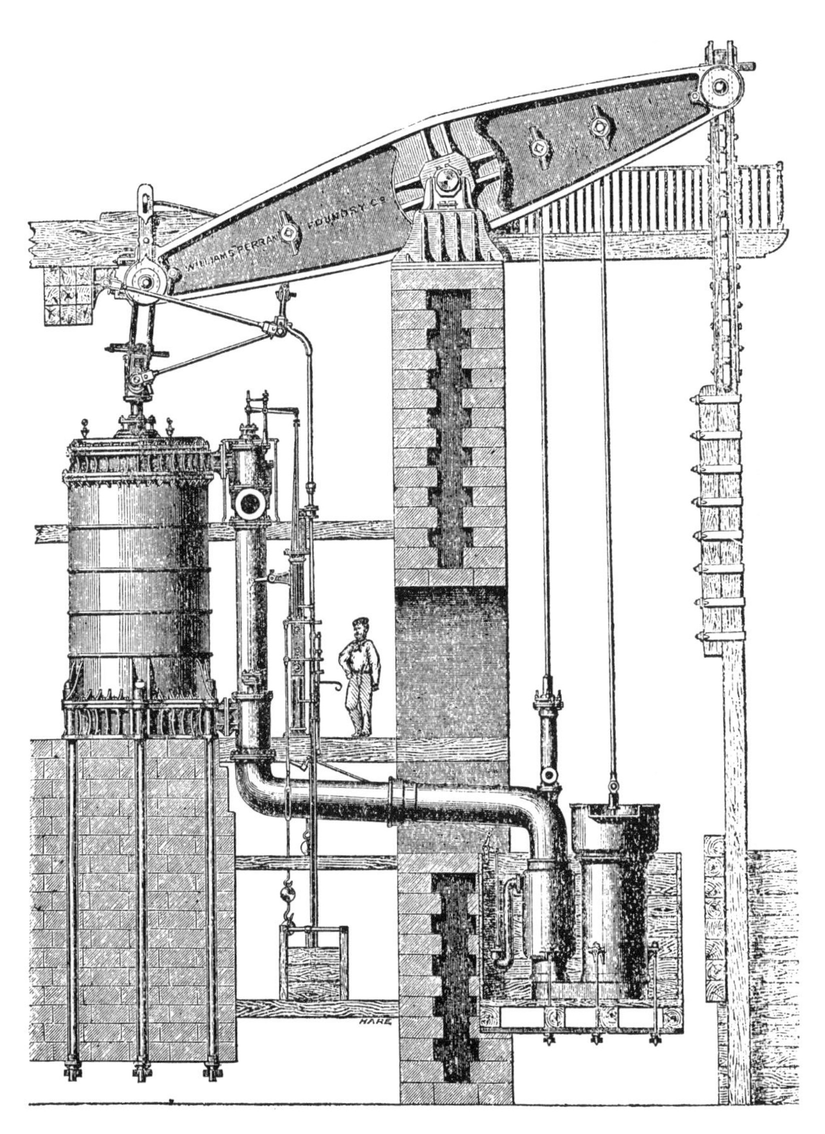

Plate 8. The eighty-inch Cornish beam engine. Reproduced by permission of D. Bradford Barton, Ltd., Truro, Cornwall, from *Cornish Beam Engines*.

Plate 9. The portal of the Sutro Tunnel, as it appeared in 1967. Photograph by the author.

CHAPTER V

The Growth of Placering

THE GREAT DAYS of cradle and sluice ended by the end of 1852 with the working out of virtually every fluvial claim. Most of the diggings were then abandoned to Mexicans, Chinese, or fate. But those men who were enamored of mining and who had held on to their stakes began to turn their attention to mining in the Mother Lode or to hydraulicking the gravel banks of the Big Blue Lead, whose reserves of gold were estimated to exceed $500 million (at then-current prices).

Hydraulicking

Hydraulicking was invented (or, properly speaking, reinvented) on March 7, 1853, at American Hill, north of Nevada City, California, by Edward E. Matteson. He was impelled to the experiment after having nearly been killed by a caving gravel bank which he had been undercutting to remove the pay streak commonly found in the lower strata of such deposits. It occurred to Matteson to wash out rather than dig out the pay streak, and to this end he consulted with Anthony Chabot, a sailmaker, who constructed the pressure hose and Eli Miller, a tinsmith, who built the sheet-iron nozzle. Chabot had already experimented with ground sluices cut by water wash and apparently contributed this idea as the gold-saving lash-up. To the delight of the three men their joint idea worked splendidly, and a new industry was born (Fig. 32).[1]

Although the per-ton cleanup in hydraulicking was not great and the investment required was considerable, the low cost of operation and the huge tonnages it could handle made the operation well worthwhile. One located a promising gravel bank and checked its gold content by booming—impounding a pond of water above the bank and then releasing the water suddenly to cut a prospecting gully through a cross

[1] Robet L. Kelley, *Gold vs. Grain: The Hydraulic Mining Controversy in California's Sacramento Valley, a Chapter in the Decline of Laissez Faire*, 28–30 (hereafter cited as *Gold vs. Grain*).

Fig. 32. Very early hydraulic mining in California. From Louis Simonin, *La vie souterrane* (1867).

Fig. 33. The headbox above hydraulic workings, showing the penstock below the flume. From Thomas Egleston, *The Metallurgy of Silver, Gold and Mercury in the United States* (1887–90).

section. If the prospects augured well, the next step was to dam a high stream and conduct its waters to the workings by flume. At the highest point overlooking the site the water went through a headbox, equipped with an adjustable overflow gate valve, into the conical top of the iron standpipe, or penstock, which ran down to the foot of the gravel bank (Fig. 33). There the water was ejected from a pivoted nozzle like that of a fireboat. That was the monitor, the direction of which was not particularly arduous work for one man but whose stream could wash out tons of sand and gravel in an hour.

The incredibly filthy backwash from the bank was led to a massive primary sluice, whose riffles were composed of flat boulders or squared sluice blocks the size of railroad crossties—even so, erosion necessitated their frequent replacement (Fig. 34). In the California system the current was then divided into a silt stream, or "undercurrent," and a grizzly stream, whose major purpose was to get rid of cobbles and shingle by straining it through a massive grid. With its more destructive components removed, this stream was recombined with the undercurrent and led to secondary sluices, where the more delicate amalgamation

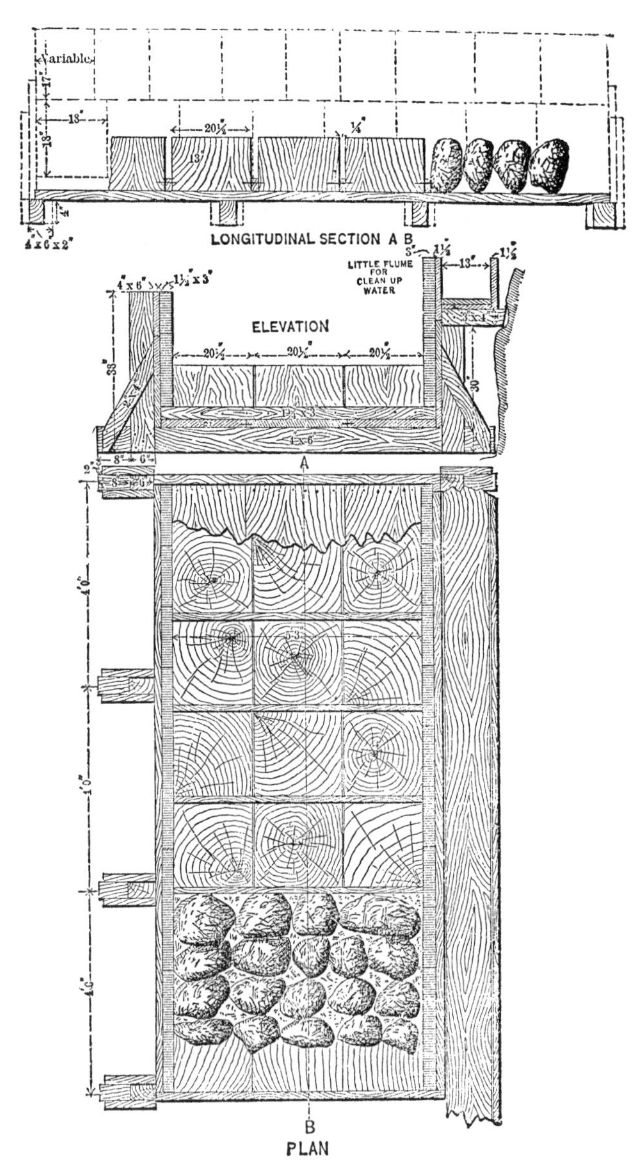

Fig. 34. Hydraulic mining, primary-sluice section. From C. Le Neve Foster, *A Text-Book of Ore and Stone Mining* (1897).

plates and catch blankets could do their work without being pounded to pieces (Fig. 35).

The more things change, the more they remain the same. Pliny has an intriguing description of Roman hydraulicking in Iberia which in major outline would do as a working model for California procedures. After erection of a substantial flume system on the brow overlooking their gravel bank, Iberian miners dug into the bank, hollowing it out into a large stope or underground cavern, whose "back" was supported by pillars of soil left in place. This series of natural arches supported the overburden, and when all was ready, the centers of the arches were cut through, collapsing the back of the stope and breaking it up almost as well as explosives could have done. Since this collapse was bound to be somewhat hard on the miners caught beneath it, an aboveground lookout kept an eye on the surface during the arch-cutting process and, when the overburden began to crack, gave a signal to evacuate the workings. The flumes above the rubble were then turned on to wash it out, and the backwash led to brush-filled ground sluices. (No mention is made of a grizzly.) Periodically the brush was collected and burned to clean up the gold dust it had accumulated, the preferred brush being a variety of wild rosemary whose natural gum caused the dust to adhere to it. Pliny goes so far as to suggest that this operation was conducted on speculation and without any advance survey to determine the recoverable gold percentages, if any, but that hardly seems reasonable in view of the large initial investment required for flume building, over and above the cost of what was probably slave labor.[2]

[2] Pliny, *Natural History*, Book XXXIII, *xxi*, 71–75. Pliny's remarks on mining are decidedly inferior to his comments on metals in trade and the arts, lending the impression that he learned of the latter by firsthand observation in the metropolis but of the former only in conversation. His description of Spanish hydraulicking sounds reasonable and practical, with the exception of the collapsing process, which seems unduly hazardous even for slave labor. Possibly the collapsing was done in a fashion well known to military engineers of the period, who were professionally accustomed to just such work in breaching city walls by mining them, supporting them from below with wooden props, then at a signal burning out the props to obtain the desired collapse. Exactly the same system could have been used in the hydraulic stopes, but since it would be impossible to calculate accurately the weight of the overburden, and hence the size and number of props required, the engineers would have had to proceed by estimate. If they had guessed wrong, the lookout could signal for a rapid evacuation of the stope during the arch-cutting process, and it may be that such a wrong guess was observed by the person who reported the incident to Pliny.

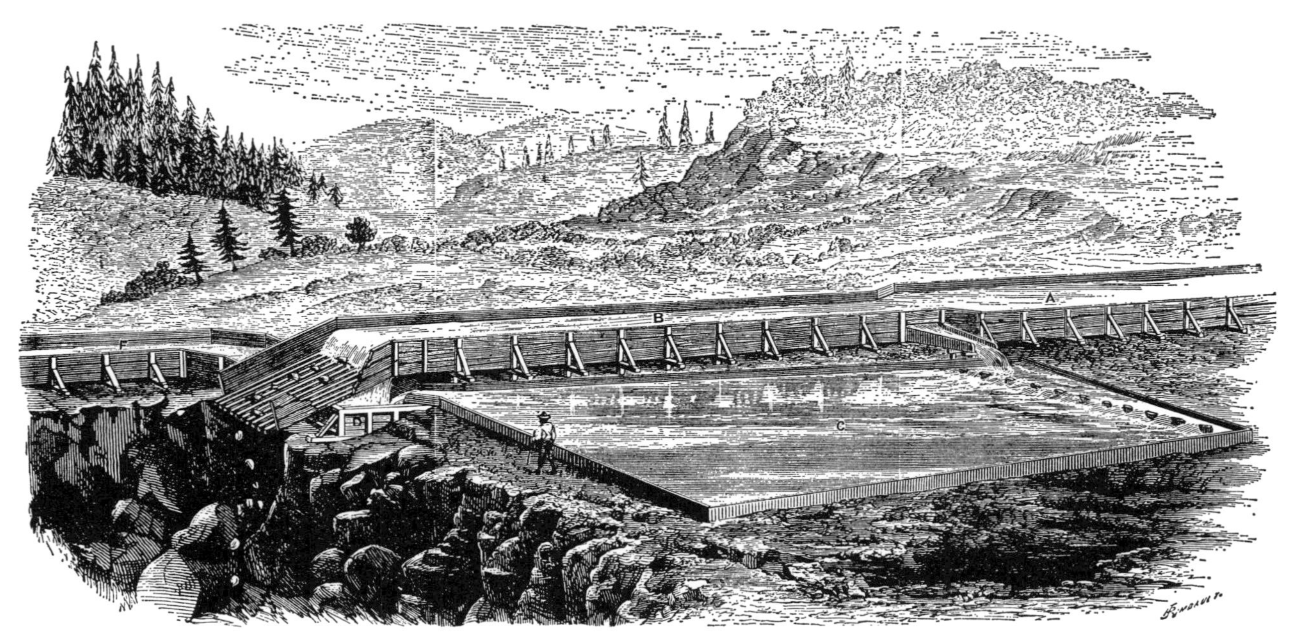

Fig. 35. Gold-saving equipment in hydraulic mining. *A*, main sluice from base of gravel bank; *B*, contracted sluice; *C*, undercurrent; *D*, drop box; *E*, grizzly at end of contracted sluice; *F*, undercurrent and drop stream recombined into continuation of main sluice. From Thomas Egleston, *The Metallurgy of Silver, Gold, and Mercury in the United States* (1887–90).

Hydraulic mining dealt effectively with remarkable quantities of low-grade gravels but had the drawback of putting into circulation vast tonnages of slickens, or sluice tailing. The easiest and cheapest way of disposing of this effluvium was to drain it into the nearest major watercourse, often by means of very expensive and extensive drainage tunnels. Because of their angular shape slickens are remarkably resistant to water scour, with the result that the interior valley river channels may be polluted and silted up. A river whose bed is rising can respond only by spreading out laterally. On this occasion the slickens deluge all but ruined agriculture in the Sacramento and San Joaquin valleys. The farmers organized antidebris associations, and in 1884 the "Sawyer decision" forcibly estopped hydraulicking except under such conditions as made it unprofitable.[3] Although the Sawyer decision professed only to enjoin the disposal of slickens in an offensive manner, the sole alternative to the ruin of an industry was the construction of impoundment dams, an approach in which no one seemed greatly interested.

During its brief heyday a technically unusual variation of hydraulicking was employed whenever the archaic gravel banks had been capped by basaltic lava flows. The procedure was to drive a vertical raise (a shaft excavated from below) upward into the heart of the deposit, bring in the hydraulic giants, and begin washing everything in sight down to the foot of the raise. There a massive grizzly shunted aside the boulders, while the undercurrents were led through a series of gold-saving sluices built right into the base of the workings. Now and then, of course, the hydraulickers would hit a mass of quartzite[4] which the water would not touch; in that event the deposit would have to be blasted out and crushed fine in stamp mills and then treated as though it were free-milling gold quartz.

[3] Cf. Kelley, *Gold v. Grain*; Thomas A. Rickard, "Farmer v. Miner," *Mining and Scientific Press*, Vol. XCVIII (April 24, 1909), 566–67; Charles S. Haley, *Gold Placers of California*, 9–12. The "Sawyer decision" was the common name for the decision in *Edwards Woodruff* v. *North Bloomfield Gravel Mining Company et al.*

[4] Quartz is of igneous, injective nature, whereas quartzite is of sedimentary origin. Quartzite usually starts as alluvial quartz sand which has been naturally cemented and consolidated into sandstone. Thereafter it is metamorphosed by heat and pressure, but its somewhat granular surface betrays its origin. Gold in such deposits would be free-milling, since it was initially deposited there by placer action.

Dredging

Placering in California remained moribund until 1898, when a novel technique was introduced at Oroville by W. P. Hammon and Thomas Couch. This new technique was dredging, which had first been reported in the New Zealand placers about 1882 and first employed in the United States at Bannack, Montana, in 1897. A hybrid of gold-rush river mining and of ordinary sand dredging, it had been developed by the New Zealanders, who had devised a floating dredge carrying its own gold-saving equipment. The dredge was not only effective but also capable of working the entire unobstructed length of river channel. Better yet, the debris of the operation went right back whence it came, producing no particular threat downstream. With appropriate modification, dredging was to prove exactly suited to California and, with more modification, to the newly prospected gold fields of Alaska.

Dredging was less effective in the foothill counties than it was in the lower stream valleys because of the radically different terrain. Hydraulicking had been best conducted in pediment highlands, where substantial differences in elevation could be utilized to provide the necessary head of water. The best dredging ground was found in contemporary river valleys whose surface and streams possessed little grade. To be sure, the gold-bearing gravels in these valleys were eroded from the deep placers of the pediment, transported thither, and redeposited, but these values suffered little in the movement. The question of low mean grade and a broad alluvial overburden was critical to dredging, since its secret lay in the creation of a broad artificial pond several feet deep, which actually moved forward with the progress of the work. It followed that any sudden increase in grade or the appearance of fall-line rock ridges across the right-of-way stream brought the work to an abrupt and permanent halt.

In California the conditions for dredging were ideal along the Folsom-Oroville axis in Placer, Yuba, Butte, and Sacramento counties. This region had been intensively worked by the argonauts, but no argonaut could work alluvium that went much less than an ounce of gold per cubic yard of pay dirt. Millions of yards of low-grade auriferous gravels were still to be found in the valleys of the American, Yuba, Feather, and Trinity rivers. And it had been demonstrated that an efficient dredge could show a profit on a gold recovery of nine cents a cubic yard.

As early as 1896, American authors had begun publishing sketches of New Zealand dredges, captioned "Nettleton Placer-Mining Machines."[5] Although the California varieties would grow enormously in size and complexity, their basic design would not vary greatly from the original version. It consisted of a large flat-bottomed barge shaped like a shoebox, built by bridge carpenters either in a pit beside the river (which would later be admitted to the excavation) or on bank-side skids for sideways launching. The most prominent feature of the dredge was the massive bow ladder, or boom, around which circled the endless bucket line which did the actual digging. Although dipper dredges, which dug with a single excavation shovel, were experimented with, the bucket-line dredge in its Risdon or "close-connected" versions was to remain the general favorite.[6] The bucket line itself ran about two giant "tumblers" or gargantuan pulley-wheels, the upper one of which was connected to the power source that moved the line, and the lower "idler" of which was down at the submerged end of the ladder (Plate 5).

Occupying the barnlike upper works were the power plant (initially steam, later shore-based electricity); screening machinery; the winches, gantries, hoists, and spuds used for moving things about; the gold-saving equipment; and the rubble stacker. Since a dredge had no conventional means of propulsion, the operator maneuvered it by paying out and hauling in steel cables whose ends were securely anchored to buried deadmen ashore. After the dredge was winched into position facing the upstream bank to be excavated, it was anchored by a spud dropped off the stern. The spud was a great square column, raised and lowered by machinery. When its pointed butt sank into the bottom of the pond, it served as a fulcrum to hold the barge in place against the reaction of the bucket line clawing away at the bank. (For some reason which the writers fail to explain, one spud was always made of steel, and the other was always of wood.) With control exercised in this fashion, it followed that the "captain" of the dredge was the winch operator who regulated its movements.

[5] Lakes, *Prospecting*, 253–54.

[6] The first-generation dredges were of the New Zealand type. California dredges appeared in two varieties, of which the earlier was the Risdon, and the later was the close-connected type. Weatherbe, in *Dredging for Gold in California*, discoursed at great length upon the differences between the two, which were relatively minute to anyone but an engineer.

Pivoting on the stern spud, its bow pulled right and left by the forward winches, the dredge in action resembled a leg-tethered elephant slowly pulling away corn on the cob with the tip of its trunk. The bucket line moved at a fair speed, discharging a loaded bucket every two seconds into the funnel-like hopper below the upper tumbler.[7] This discharge was assisted by heavy jets of water at high pressure which played into the bucket as it swung over into the dumping position. As few as three or as many as eighteen cubic feet of semiliquid muck then gushed through the grizzly and, freed of boulders, fell into the screening machinery. Although massive shaking screens were tried, the favorite screening device was the cylindrical trommel, a large, down-slanting tube of boiler iron pierced with innumerable holes one-half inch in diameter. The gravels entered the upper end and, as the trommel slowly revolved, the finer material, including the gold, was washed out through the holes. Cobbles and oversize gravel slid down to the rear lip to drop out upon the stacker belt. The stacker was a light boom carrying an endless rubber belt which conveyed the oversize to its final resting place at the back bank of the dredging pond.

The fines which dribbled out of the trommel holes were collected in a launder and conducted down to the gold-saving machinery which occupied the interior spaces at the right and the left of the trommel. Save in size and elaboration, this gold-saving machinery differed in no important respect from the sluice box of the argonaut. For riffle bars it employed steel angle irons holding mercury pools as gold traps. Instead of the forty-niners' Brussels carpet, it employed coca matting to entangle the finer colors. In the weekly cleanup ritual the riffle bars were pulled out and washed down, the mercury traps were scraped out with a spatula, and the coca matting was rinsed in tubs of water. The several gallons of gold amalgam and black sand so collected were then separated in the classical hand-operated rocker. Engineers shook their heads at such arrangements, pointing out that only "dimensional" gold colors were recovered, that a great deal of flour gold got away, and that the limited capacity of the gold-saving equipment restricted the yardage the dredge could excavate daily. Apparently these professional criticisms failed to impress the conservative managers of the operations.

Except for down time necessary for cleanup and repair, the dredge

[7] George J. Young, *Elements of Mining*, 474.

unceasingly chewed its way ahead, following the right-of-way stream valley. The valley had been carefully prospected by means of a well-drilling rig, and the colors so obtained had been rigorously collected and counted and then plotted with extreme care for determination of profitable ground. Ahead of the pond went gangs of Chinese, clearing and stumping the land. Despite its size and power, the dredge had a very sensitive digestion, with a particularly dyspeptic reaction to wood in any substantial amount. Embedded tree trunks were especially troublesome, and otherwise promising terrain was often passed over when the prospecting rig revealed too many buried snags. Almost as aggravating were the abandoned and infilled shafts, coyote drifts, and other workings of the argonauts who had placered the land half a century before. Surprisingly enough, boulders represented no particular problem; an oversized boulder could not be retained in the bucket, and would roll off to one side when nudged out of the bank.

As the dredge advanced, the pond advanced with it, excavated in front with the bucket line and infilled behind with the spoil of the work. Across this filled ground utter desolation prevailed. The silts discharged off the stern from the gold-saving equipment settled first to the bottom of the pond. The cobbles and oversize were then dropped on top of this silt bank by the stacker. The result was a vast reach of irregular, shingly wasteland which was unfit for any purpose but eucalyptus planting. Even in that haphazard era engineers were faintly uneasy about the despoilment of the landscape. It was obvious that if the cobbles were dropped first and the silts somehow pumped over them, the effect would be far better, but that suggestion was hardly more than mentioned when it was quietly dropped. About all that could be said for dredging from the conservationist's viewpoint was that the silts at least were firmly anchored in place, not turned loose to menace the country downstream.

The technical growth of California dredging in the decade 1897 to 1907 was remarkable. The first Hammon and Couch machine was so small and primitive that its bucket-line link pins were ordinary 0.5-inch carriage bolts and its bucket capacity was little more than 1 cubic foot. Within ten years the Folsom Development Company was constructing giant steel dredges whose bucket link pins were 4.5-inch alloy-steel bars connecting massive buckets of 13-cubic-foot capacity. Horsepower and yardage capabilities increased proportionately. The art itself advanced from an eccentric branch of placering to an engineering specialty warranting its own authorities and textbooks.

The high degree of visibility and the concentration of production of the industry also made the garnering of its statistics relatively easy. It was estimated by Weatherbe that net monthly profits from each dredge ranged from $2,000 to $5,000, depending upon the ground and the efficiency (usually equated with size) of the equipment. By 1909 the total investment in dredging equipment in California amounted to $40.6 million, and the annual placer-gold production reported in the same year was $8.7 million.[8] Thus the average monthly gross production of each dredge was about $14,500, and profits on the investment varied from 14 to 40 per cent, provided the production of placer gold by other means was negligible, as it probably was.

The boom deflated almost as swiftly as it had begun. The decline had become visible by 1919. In that year the total number of dredging operations was reduced from 687 to 112. Although much of the fall-off occurred in the Alaskan mining grounds, California undoubtedly lost its proportional share. The state's placer-gold output also fell off from the peak of $8.7 million to $7.4 million, the bulk of which was produced by a handful (perhaps twenty) of the largest and most effective dredges.[9] California's third placering boom was over; there would not be a fourth.

Concentration

Although associated with milling, concentration was an offshoot of placering. Its purpose was the separation of heavy mineral from gangue (usually low-grade sand or mill screenings) by gravity and inertial separation in a water medium. In milling the detritus was crushed rock particles; the desired residue, when not native metal, was a metallic sulphide. Even though the principle was the same, the equipment was not identical, chiefly because of the fact that concentration dealt with more finely crushed material, and frequently because of the need to conserve water. To offset these problems, mechanical power was usually available. It can be generalized that all frontier milling involved con-

[8] U.S. Bureau of the Census, *Thirteenth Census of the United States: Abstract . . . with Supplement for California*, 572, 677, 685, 720.

[9] U.S. Bureau of the Census, *Abstract of the Fourteenth Census of the United States*, 1268, 1272, 1281, 1284.

centration, but that mine-head or mineral-district milling frequently stopped there as well, owing to lack of ore, water, or fuel essential to carry the process through the more desirable reduction to matte, or desulphurized mixed crude metal (Fig. 36).

Before gold-rush times, concentration was usually carried out by means of equipment which relied upon the sound old buddle principle. Bauer devoted much space and many woodcuts to its variations, some of which appear to have been merely experimental lash-ups and even simply proposed alternatives. Buddling accorded ill with the Cornish-American practice of get-in-and-get-out, wherefore frontier concentration rigs speeded up the slow washing process by shaking the equipment, usually laterally to the water flow. Correctly executed, this process multiplied the separation effectiveness by adding the factor of inertia to that of specific gravity: sulphides, being the more massive, would not be moved as far as would the lighter gangue particles. It may be that this trend originated in the observation that the placering rocker provided the most effective separation; indeed, as late as 1920 the rocker was still in use as the primary cleanup device. For that matter, at the Golden Belt Mill in Arizona, I observed an ingenious final-stage concentrator which was little more than corrugated culvert sections shaken laterally by an electric motor.

The first frontier concentrator as such may have been the cistern jig, a sieve-bottomed frame which was suspended half in and half above the water level in a shallow cistern. It seems to have originated in the Ozarkian galena mines, where it was actuated by a sort of seesaw, on one end of which was the jig, and on the other the operator, who jounced the jig about in a manner similar to the Mexican *trapiche*. In California times the jig was hung on flexible straps and joggled about either by hand or by a power-driven eccentric rod. The operator, armed with a wooden board or paddle, stood over the affair as roughly screened crushed ore was poured in and the jiggling begun. Light, valueless pieces of larger size were jigged to the surface of the charge, whence they were swept away with a sideways flick of the paddle board. After several charges had been sorted in this manner, the heavy accumulation in the bed of the jig was dumped into a receiver, where it was presently joined by the fines scooped up from the bottom of the cistern. It was not a very effective device, and apparently was soon abandoned.

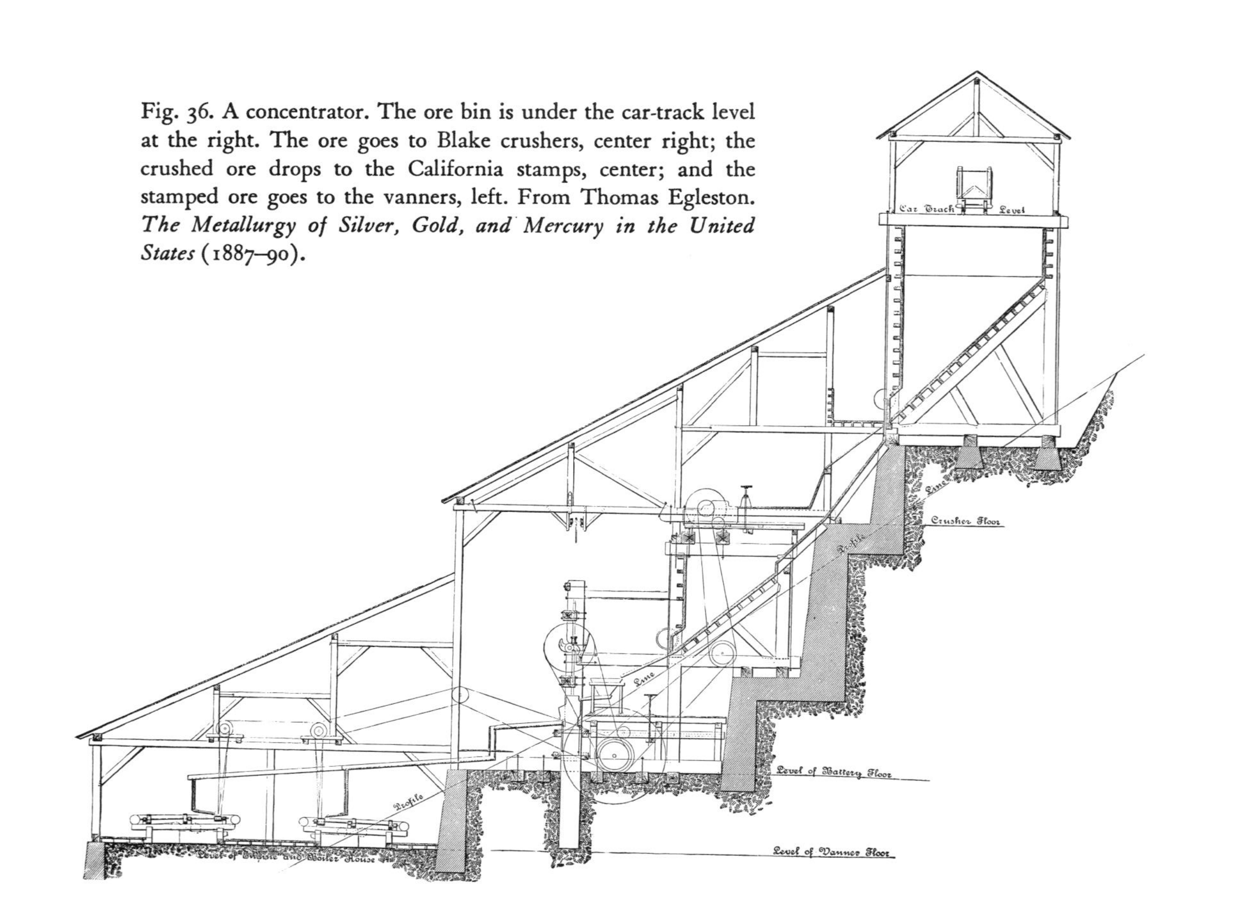

Fig. 36. A concentrator. The ore bin is under the car-track level at the right. The ore goes to Blake crushers, center right; the crushed ore drops to the California stamps, center; and the stamped ore goes to the vanners, left. From Thomas Egleston. *The Metallurgy of Silver, Gold, and Mercury in the United States* (1887–90).

What effectiveness the jig possessed rested less on its motion, the work of the operator, and the immersion in water than in the significant fact that metallic sulphides are usually much more friable than the gangue in which they are found. That means that if roughly crushed ore was closely examined it would be found that the dusts and fines contained a much higher percentage of values than the over-all assay returns of the ore. Contrariwise, the fraction most resistant to crushing would hold far less than its arithmetical share of values. That being the case, an effective upgrading could be obtained by screening such roughly crushed ore and recovering the fines. With such an ore as galena, this elementary form of concentration could upgrade 5 per cent ore to 50 per cent concentrates with but one pass through the screen.

It was soon discovered that for wet concentration a much better device than the jig was the vanner, whose name was derived from the Cornish verb meaning "to upgrade ore by washing out dust on a shovel blade." In its original form the vanner consisted of a broad, endless canvas belt, slightly inclined laterally. Along the high side was a perforated water pipe, carefully adjusted so that it jetted gentle streams of water across the surface of the belt. Mineralized screenings were allowed to fall in a thin stream onto one end of the belt, and, as the belt moved, the water washed away the gangue particles, which went over the lower side. Heavier particles stayed on the belt to the turn of the far roller, where they dropped or were scraped into a hopper. The similarity in principle of a vanner to a buddle, or even to the planilla, is apparent.

An inventive Irishman, William B. Frue decided to improve the vanner by (figuratively) taking it apart and reassembling it sideways, upside down, and backward. For canvas belting he substituted rubber. He abolished the lateral slant, making it axial. He placed the water jets *across* the belt and added a sideways shaking mechanism to the chassis. A gear device enabled the operator to change the over-all angle of axial slant as desired. When finished, the Frue vanner was wonderful to behold, and as such was advertised all over the West. It must have been a technical success, reducing appreciably the tonnages that had to be handled farther along in the mill or concentrator by more expensive processes. Certainly it linked Frue's name as inextricably with his miraculous vanner as Henry Ford's would be with the model T (Fig. 37).

Fig. 37. The Frue vanner. From Joseph C. F. Johnson, *Getting Gold* (1917).

The Northern Rockies Gold Rush

Certain now that gold was to be found in the West, and having perfected their technique of finding and capturing it (they thought), prospectors began to fan out in all directions from California. The action began in the intermountain valley between the Bitterroot and Jefferson ranges of the northern Rockies where, about 1852, a Red River métis, François Finlay, alias Benetsee, found some gold nuggets at Deer Lodge Valley. The news was disseminated among the retired beaver trappers who resided with their Indian women in the vicinity, but nothing particular happened until the enterprising Stuart brothers, James and Granville, drifted into the region from California in 1857. The Stuarts began to prospect about the Hellgate Fork of the Bitterroot, finding colors and building a sluice to garner more. They were joined by other vagrant prospectors, who helped them found the optimistically named Gold Creek Camp. Despite the camp's name, the Gold Creek sands at first gave up only three dollars in a hundred panfuls, but by the next autumn a modest quartz lode had been located. Its owners worked it with a water mill running some sort of crusher. Greever says that the crusher was a stamp mill,[10] but the fact that it was adapted from wagon wheels leads to the suspicion that it was really a Chilean mill, with stone weights lashed to the spokes to increase its weight.

In 1860 victims of the Fraser River rush and bust were placering on

[10] William S. Greever, *The Bonanza West: The Story of Western Mining Rushes, 1848–1900*, 217 (hereafter cited as *The Bonanza West*).

the other side of the Bitterroots from the Stuarts, just about making grub but little more. None of the prospectors involved appears to have reflected that finding gold on *both* sides of the same range strongly suggested a pronounced deviation from the Sierra Nevada pattern on which they had been trained. True to tradition, however, the interior valley prospectors continued to work south and up the tributaries of the Hellgate Fork. On July 28, 1862, they got their reward: the John White party found substantial colors on Grasshopper Creek in the Beaverhead Valley. This find precipitated a real rush, the boom creating the camps of Virginia (of the North), Bannack, and Alder Gulch. Population figures were given impetus by the simultaneous collapse of the Confederacy in the West, many an erstwhile secessionist finding it convenient to seek his fortune in Confederate Gulch, which, despite its name, did not extend a welcome to Confederate conscription agents. At the same time the great purge of the hoodlum element in 1855 by the San Francisco vigilantes sent a wave of health seekers to the more salubrious climates of the Montana gold camps. Between the two elements Virginia and Bannack were turned into hells, and miners who attempted to leave with their cleanups were dry-gulched and robbed. Before long a local vigilance committee headed by the formidable J. X. Beidler was dealing with the health seekers in quite effective fashion—at least none of the subjects of Beidler's attentions were heard later to complain of their treatment.

The Montana placer rush started prospectors fanning out widely, and they promptly discovered hundreds of lodes. Mills opened for business, and Montana Territory boomed. In 1864, Butte was discovered by Budd Parker, P. Allison, and Joseph and James Esler (or Esher). They found placer gold which turned to silver in a manner exactly like that of the big strike of the Comstock Lode, but it then turned to copper, like the United Verde. The irony of it was that the prospectors kept working upstream when they should have wandered down. It was not until comparatively late in 1885 that an unemployed carpenter named Noah S. Kellogg obtained a slender grubstake of seventeen dollars and a jackass and went poking about the Coeur d'Alene district, where a short-lived placer district had played out. Kellogg had more luck than he deserved. He struck two rich silver-lead-zinc lodes, which he staked as the Bunker Hill and the Sullivan. He defaulted on his grubstakers, and there was a series of lawsuits which ended with syndicates gaining control of his two ledges, as well as a third, the Sunshine.[11]

Kellogg's discovery has been preserved in mining-camp legend as a typical example of what may be called, for want of a more pungent expression, "burro dowsing." The presumed ability of jackasses to find mineralization was as much touted in Anglo circles as were the prospecting talents of ghosts in Hispanic quarters. Jim Wardner, a promoter of the area, put the tale of Kellogg's jackass in glowing prose:

> Looking across the creek we saw the jack standing upon the side of the hill, and apparently gazing intently across the canyon at some object which attracted his attention. We went up the slope after him, expecting that, as usual, he would give us a hard chase; but he never moved as we approached. His ears were set forward, his eyes were fixed upon some object, and he seemed wholly absorbed. Reaching his side, we were astounded to find the jackass standing upon a great outcropping of mineralized vein-matter and looking in apparent amazement at the marvelous ore-shoot across the canyon, which then, as you now see it, was reflecting the sun's rays like a mirror.[12]

In commenting upon this tale, Rickard gave vent to cynical amusement. He pointed out that argentiferous galena exposed to the atmosphere is dull in color, a matte black or rusty red. He concluded acidly, "The talk of a glittering mass of silvery ore sticking out of the mountain-side so brilliantly as to mesmerize the ass, *and others not any wiser* is pure moonshine."[13]

Early Days in Arizona

At about the same time as the Montana strikes acquisition of the Spanish Southwest by the United States led to revived interest in the mineral wealth of Arizona, so long reported and so long ignored. A few Americans began to drift in, seeking a fast dollar in mining. By the canons of strict logic their activity should have been confined to placering along the Colorado River. In fact, they tended to stay south of the Gila River, the chief highway through the territory, and to interest themselves in

[11] These mines are still active, as the results of lava capping, which preserved them from the usual fate of supergene enrichment above and consequent dead *borrasca* below.

[12] Rickard, *A History of American Mining*, 321–22.

[13] *Ibid.*, 322. Italics added.

the silver lodes about Tucson and Yuma. Conditions in the territory were not salubrious. The Arizona of the period has been described as

> a delightful country in every respect, except in climate, soil, production, and inhabitants. The natives have a pleasant way of slaughtering every stranger who attempts to stay there; and sometimes, when they refrain from their amusement for a few months, the strangers fall to killing each other. Until very recently it was said that no white man had ever died in Arizona with his boots off, —meaning that he had never died in bed. The Indians make traveling very insecure; and the Peons, or native Mexican laborers, in the mines vary the monotony of their employment by an occasional massacre of the superintendent and every other white man about the place.[14]

Ajo, known to the Jesuits, was promptly relocated by Peter R. Brady, and its silver was skimmed off by a San Francisco syndicate, the American Mining and Trading Company, which sent its gleanings out by way of the Colorado River, manifested to the smelters at Swansea, Wales. Two other promoter-adventurers, Charles D. Poston and Herman Ehrenberg, prospected about the miserable hamlet of Tucson. They found silver chlorides in the Santa Rita Mountains south of the place but ultimately gave them up for much the same reasons that had discouraged the early Spaniards. It is worth noting that most of Poston's and Ehrenberg's associates were sooner or later killed by Apaches, thus dying in what was, for Arizona Territory, a natural fashion.

Another devotee to the main chance in this place and period was Lieutenant Sylvester Mowry, individualist and lecher, who went to the region originally on Army business but in 1860 found time from his other occupations to acquire title to an ore deposit about twelve miles south of the present settlement of Patagonia, on the south side of the Santa Ritas. Mowry did better than Poston and Ehrenberg, for his ore was amenable to direct smelting. He poured his own bullion and for several years prospered in his hobbies, but with the coming of the Civil War he unwisely let himself fall under suspicion of harboring Confederate sentiments. When Union troops entered Arizona from California, Mowry was put into the *juzgado* for the duration and then expelled from the territory.

Without detracting from the nerve and enterprise of any of these

[14] Thomas W. Knox, *Underground; or, Life Below the Surface*, 699–700.

men, I find it impossible to avoid the conclusion that they were chloriders—in the grand manner, perhaps, but still chloriders. None save Mowry appear to have gone deeply enough to encounter sulphurets but merely confined themselves to working surface outcrops by open-pit methods. The short haul from Ajo to Yuma Crossing may have enabled the American Mining and Trading Company to break out with a profit; Mowry paid expenses with bar bullion; but Poston and Ehrenberg had two pieces of ill luck. Lacking water for concentrating or fuel for smelting, they had to hand-sort and cob their ore by hand. The ore thus up graded then had to be hauled several hundred miles to Yuma or Guaymas across the desert at costs which undoubtedly ate the partnership alive.

Large-scale placering in Arizona began in 1862, when a party of Californians under the scout Joseph Reddeford Walker made a long, roundabout march across Nevada, Colorado, and New Mexico, falling in with the Gila headwaters and following that river to its big bend. There they turned north on the Hassayampa River, following it up into the Bradshaw Range of the central mountain belt. The hearts of the Walker party must have quickened as they rode, for the country is brown Precambrian schist below, a light-gray dioritic Tertiary granite above, with quartz float covering the ground, and white quartz dikes and croppings everywhere protruding from the earth in such profusion as to resemble melting snowbanks. The men panned the watercourses, finding placer color in attractive measure. Within two years the mountain placer camps of Lynx Creek and Bradshaw City were booming. The prospectors ultimately located the lodes of mineral origin intruded between the schist and the diorites. Although numerous, the lodes were never greatly profitable; the lack of water for concentrating (the rule-of-thumb requirement is eight tons of water to one ton of ore) and the inordinate costs of transportation in those rugged mountains prevented this district, Crown King, from becoming more than a bait for promoters with an eye to the main chance.

In the meantime, the Austrian-born prospector Henry Wickenburg and three companions were investigating the Bradshaw pediment to the southwest, working about the base of the volcanic Vulture Mountains, which are really igneous dikes. On the flat Wickenburg found a quartz outcrop large enough to be deemed a butte. Virtually all of the butte held some gold, although the profitable deposits were in four shoots, with high-grade concentrations occasionally found within the shoots. The country rock and gangue of the Vulture Mine are a colorful and incredibly complex mélange ranging from schist and shale to andesite, granite, porphyry, and quartz. The outcrop was not mined but quarried; Wickenburg found the best concentrations of gold, packed them to the Hassayampa, and there ran them through an arrastra. After a while he found it more profitable to begin leasing, selling the ore in place at fifteen dollars a ton, leaving it up to the purchaser to mine, concentrate, and pack the concentrates to the Colorado River. So many labored there that they ultimately left a substantial glory hole which remains visible to this day. But few apart from the forty thieves who high-graded its ore made much profit, owing to scarcity of water and cheap transportation.

About the same time (the 1860's) Captain John Moss, stationed at Fort Mohave on the Colorado River, located a rich pocket of gold in the Black Mountains which canalize the river. It is doubtful that Moss realized that the Black Mountains mark the sides of—and the river itself follows—a major fault-fissure. Moss's lack of knowledge of geology did not prevent him from making off with an estimated two hundred thousand dollars in high-grade surface ore, a nice supplement to a soldier's pay, before selling the picked bones to an eastern syndicate, which promptly lost its collective shirt trying to make something of what was left of the Moss Mine.

A hard case, "Colonel" Jacob Snively, led a party up the Colorado River to wash its sands, finding some color and comparatively good transportation up to the future port of Ehrenberg. The Colorado gold was deceptive, however. Some of it may have come from the Moss Mine and the Oatman District lodes and the rest from river-bed outcrops of the fault, but the steep straight run of the river washed away most of the gold that it did not hide from sight, so as accessible sands were riffled out, the inhabitants drifted away. Throughout most of the West, in fact, placering had played out as the major gold producer by the end of Civil War times. Thereafter the emphasis would be upon lode mining.

CHAPTER VI

Colorado Mining Days

The gold and silver mining frontier reversed the usual pattern of American migration; it was an eastward, not a westward, movement. Virtually all its fellows, whether coal, timber, iron, or petroleum, were prospected and developed by men who marched toward the setting sun. The gold seekers, on the other hand, jumped to the Pacific and then recoiled by means of a backwash of adventurers indoctrinated in the arts of placer prospecting. They began, reasonably enough, by searching for the Mother Lode. They found it—or the quartz dike that passed for it—and then went north to the Cowlitz and Willamette valleys, whose topography was superficially similar to that of California. Finding little enough in Oregon or along the Fraser River of British Columbia, the prospectors drifted southeast to the Montana placer camps to start the northern Rockies boom. Then, in 1858, came news of placer color in Colorado.

The Pikes Peak Gold Rush

It began in 1850 when a California-bound party of Cherokee Indians from the Indian Territory, including among their number a Captain John Beck, was proceeding westward by way of the South Platte Fork of the Overland Trail. The Cherokees' original home was Georgia, and there they had apparently picked up the rudiments of placer panning, for in the streams near the site of the future Denver they found some gold color. The showing was not enough to detain them, but Beck was sufficiently impressed to return to Georgia eight years later, recruit William Green Russell, who was experienced at placering, and return to the site of the original discovery. Great minds run in the same channels: in Missouri and in Kansas Territory other parties were being formed for the same purpose. Most of these prospectors, including Russell, found color only in moderate quantities about the South Platte tributaries, but the newspapers exaggerated the finds to such an extent that the "Pikes Peak or bust" rush of 1858 brought a flood of boomers to what would become Colorado Territory.

The rush was deflated in 1859; of the one hundred thousand men who had entered Colorado the previous year, only one-fourth remained to pick over the gravel of Cherry Creek for meager returns. Then, on May 6, 1859, matters dramatically took an unexpected turn for the better. One of the boomers was a Georgian, John H. Gregory, who had left his family with some intention of gaining his fortune in the Fraser River rush. Winter-bound at Fort Laramie on the Overland Trail, Gregory had spent his idle months panning southward between the Cache la Poudre and Pikes Peak, then working up the Vasquez Fork (Clear Creek) of the South Platte. He took whichever fork of Clear Creek showed the best indications, coming at last to a side ravine—now Gregory Gulch—where the color suddenly increased and then abruptly faded out a bit higher up. Gregory suspected that he had just passed the outcrop of the lode of origin, but was interrupted at this point by a spring snowstorm and a shortage of food.

Gregory returned to his base, Golden, where he borrowed a grubstake from David K. Wall, a veteran forty-niner, and enlisted two partners, Wilkes Defrees and William Ziegler. Early in May, 1859, the three associates returned to the promising ravine. Gregory set Defrees and Ziegler to digging prospect holes about the foot of the steep "nose" which marked the junction of the ravine and Clear Creek, while he himself scrambled about on the heights above. Whether he realized it or not, he was soon standing atop a massive quartz dike which dipped straight down the point of the nose. Scooping up a panful of surface dirt, Gregory slid down the slope to wash it out. The drag held a half ounce of gold, a phenomenal return.

Gregory spent the night of May 6–7 in wild surmise and at daybreak began searching along the side ravine for the lode. He found it and ordered Defrees to start digging, and the two men soon uncovered fragments of quartz displaying the typical rusty, spongy appearance of promising float. Gregory took a panful of the float, crushed and washed it out, recovering between one and two ounces of gold. On May 10, certain of his find, Gregory staked two claims (including the bonus for discovery) and began trumpeting the news. A massive stampede began, on the strength of which Gregory sold out for twenty-one thousand dollars, and was retained at a salary of two hundred dollars a day to search for other lodes on the account of the purchasing syndicate. It was not long before he discovered a major side shoot of the original dike. Immediately three camps sprang up in the district: Gregory Gulch, which became the site of Mountain City; the north fork of Clear Creek at the foot of the lode, which became Black Hawk, the future milling and smelting center; and, higher up Gregory's ravine, Central City, in those "mountain diggings" (a technical solecism which reversed the more common malapropism placer mines). The lode was worked by open-cast methods, leaving a great artificial cut in the nose of the promontory which remains clearly visible to this day.[1]

In solving one problem, Gregory had created dozens of others. To begin with, his strike and those which soon followed it were not placers but lodes, and no one was at all certain how to go about extracting gold from hard rock. Prospectors reared in the California tradition were puzzled by the fact that the Colorado lodes did not at all resemble the Mother Lode formations but seemed instead to crop out radially from the peaks every which way. Erosion and weathering had left a great deal of free gold at the surface of the outcrops, gold which could be recovered by simple crushing and concentration, but the deeper the prospectors delved the faster the gold recovery seemed to fall off. The Coloradans buzzed about like bees, recognizing that wealth lay beneath their feet but baffled by the difficulties of getting to it.

Much later it was deduced that during Tertiary times the entire mid-continent region may have been submerged beneath an ocean of considerable proportions. At that time there had occurred eruptive double fissuring of the crust where the Rockies now stand (hence the parallel twin ranges). The overlying cold and counterpressure of the ocean had prevented the acidic magma, then upwelling, from breaking through the surface and distributing itself in fine dispersion. Instead, the magma had been contained, chilled, and required to spread out beneath the crustal strata, where it had formed a double row of enormous batholiths, in cross section rather resembling mushrooms perched on the slender

[1] The primary source for Gregory's strike is Ovando J. Hollister, *The Mines of Colorado*. Another version, including photographs of the gulch, is to be found in Robert L. Brown, *Ghost Towns of the Colorado Rockies*, 235–41. Somewhat puzzling are the sarcastic reflections on Gregory, presumably extracted from Hollister, given in Bancroft, *History of Nevada, Colorado, and Wyoming*, 337 note 31.

stems of their vents. In the following Quaternary period the mid-continent region had been uplifted, and the sedimentary overburden had been swiftly eroded away, leaving the granitic domes exposed—hence the considerable differences between the domed Rockies on the one hand and the tilted blocks of the Sierras on the other.[2] When viewed from the air, the symmetrically rounded summits of the Front Range and the uptilted circular strata which surround discrete domes bear out this contention with remarkable clarity.

The quartz shoots which branched out radially through these domes held a great deal of heterogenous mineralization, with the proportion of free-milling gold increasing as the lead approached the surface. Most often this gold was found only as minute specks disseminated widely through the lode or the contact zone surrounding it. Occasionally it was concentrated into "wire" or "leaf" gold of great attractiveness. In the Cresson Mine near Cripple Creek the Colorado prospectors hit the jackpot—a fabulous vug or geode within an old vent. In a cavity so large that a man could stand upright within was deposited $1,200,000 in completely pure gold crystals. All that was required was its removal and casting into bullion bricks.

In most of the lodes, particularly at depth, the gold was telluritic and carried by pyrites of iron, copper, and arsenic. Natural reduction of such mixtures had taken place as erosion had brought the pyrite nearer the surface. The sulphur had leached out, leaving, for example, iron as hematite in a sort of amorphous mass, half filling the cavity in the quartz, which often retained the shape of the original pyrite cube. There in a nest of spongy brown iron oxide might rest a handsome piece of native gold, which further erosion would release (with its partners) to form such a rich placer as Gregory had discovered. Iron or copper blossom would help mark the outcrop, thus providing a second indication of its existence. While good for the prospector—unless rounded, black iron pebbles which they called slickers were about, some old-time prospectors considered it not worthwhile to wash even one panful of sand[3]—pyrite ores boded no good for the millmen.

[2] Fred M. Bullard, *Volcanoes in History, in Theory, in Eruption*, 43.

[3] Lakes, *Prospecting*, 14. For every rule in prospecting there seems to be at least one major exception. In this case, the fact that one iron salt, limonite, frequently found in conjunction with gold is yellow represents a contradiction to the earlier

On the positive side, the early Colorado workings were quarried or drifted into firm granitic country rock which required minimal support and timbering. In addition, the lodes were so far above the mean water table that the mines were dry or at worst leaked only moderately from surface accumulations. Elsewhere water was to destroy mines and syndicates without number. Apart from flooding, water comes as close as anything to the alchemical conception of the universal solvent, or alkahest; given sufficient time, water under heat and pressure, or working in conjunction with the atmosphere, *is* a universal solvent. A dry mine may have persistence of ore to indefinite depth and its engineers can plan accordingly, but a wet mine is in trouble from the first day water is encountered. The erection of a pumping and drainage system is extremely costly. Below the water table the ore may thin out to the point where it becomes unprofitable to break out and hoist. The combination of flooding and *borrasca* ore was the ruin both of the Comstock Lode and of Tombstone.[4] In view of their other difficulties, perhaps it was just as well that the Coloradans were spared this obstacle.

In dealing with the telluritic lodes, the Americans were at first at a loss to proceed by any other than the antiquated Spanish methods they found being employed around Durango, Pinos Altos, and Gila City. The Spanish system had the advantages of requiring little capital and no formal education, but it or its steam-powered elaborations were virtually useless for the refractory ores of the Central City District. It was equally apparent that the Spanish ways of the Mexican *barretero* or the methods of Allegheny coal miners also left something to be desired. The owners addressed themselves to the problem by importing thousands

---

generalization that yellow is an unattractive hue to the prospector. To give one example, Goldfield, Nevada, waste is as yellow as a canary from limonite.

[4] Silver in the primary ores of such desert deposits as Tombstone was not very valuable, typically running about five ounces a ton. The area above that was the area of secondary enrichment, and in Tombstone it was good but not really superlative either in tenor or in quality. In my opinion there is no silver ore at Tombstone below the five-hundred-foot level (the normal water table) except for one little patch of thirty- to forty-ounce ore I have seen; all the silver went except for what was missed above the water. Of course, there may be lead-zinc-copper below five hundred feet, but Phelps Dodge had the entire camp for years, and let it go. On the other hand, in the Sunshine at Kellogg the ore is still holding well below the three-thousand-foot level, but the miners have not yet hit the water table. It is a dry mine, surface water being cut off by lava capping (Robert Lenon).

of Cornishmen. The "Cousin Jacks" could drill, blast, and timber, and their "captains" could even bring some empirical experience to the solution of pinched-out lodes and other subterranean mysteries. A captain was the equivalent in rank and pay to the later superintendent-engineer. Under him in the chain of command were the shift bosses, or shifters, also Cornishmen. Rickard noted that these Cornish shifters "knew better than anyone how to break rock, how to timber bad ground, and how to make the other fellow shovel it, tram it, and hoist it."[5] The saga of the Cornish miner in America has been related elsewhere, but to this day children in mining camps still repeat the couplet

> By Tre-, Pol-, Pen-, and -o
> The Cornishmen you come to know.

Their presence soon produced a somewhat uniform pattern in the western mines of the frontier period, a pattern which still largely prevails in drift mines, and which can be discussed in terms of the arts of drilling, blasting, timbering, and transportation.

Mine Operations Present and Past

The first view of a mining operation encompasses the headworks, where are concentrated the offices, living quarters (if no town is adjacent), ore bins, shops, change rooms, and storage yards for such items as timber and machinery. In the case of a tunnel mine no hoist is required; all

[5] Rickard, *A History of American Mining*, 24. The Cornishmen were in great demand because of their experience in the stannaries (tin lodes) of Great Britain, which had been worked time out of mind. The Cornishmen first began to emigrate to the United States when the lead mines about Galena, Illinois, were opened, the geological conditions in the eastern (Illinoisian) end of the lead-mineralized Ozarks approximating those of Cornwall. The Cornishmen were everywhere referred to as Cousin Jacks. The origin of the term is in dispute, but one version has it that John was as common a Cornish name as Patrick was among Irish immigrants and that whenever a mine superintendent announced his intention to hire more hands every man of the crew volunteered to send for his cousin Jack. To the West the Cornishmen brought a great thirst for whisky and an equally great hunger for "pasties," a sort of meat pie upon which the Cornishmen traditionally subsisted. The major sources for the section on mining are Joseph R. Boldt, Jr., *The Winning of Nickel*; Frank A. Crampton, *Deep Enough: A Working Stiff in the Western Mine Camps*; Dunning and Peplow, *Rock to Riches*; C. Le Neve Foster, *A Text-Book of Ore and Stone Mining;* Lewis, *Elements of Mining*; Thomas A. Rickard, *Man and Metals: A History of Mining in Relation to the Development of Civilization*; Louis Simonin, *Mines and Miners; or, Underground Life*; and Bohuslav Stoces, *Introduction to Mining.*

that is visible is the portal (entrance) with the waste dump on prominent display before or beside it. Before the advent of steam power a shaft mine was surmounted by a windlass or whim (Fig. 38). Its modern-day counterpart is the headframe,[6] a heavily braced right triangle, between

Fig. 38. The windlass. Reproduced by permission of Buck O'Donnell and Shaft and Development Machines, Inc.

[6] Those who would casually explore abandoned workings are earnestly warned not to do so—not even stand on the shaft collar to peer down, for shaft collars may be rotten, and should be avoided unless one does not mind visiting the sump somewhat sooner than one has bargained for. In addition, such mines are full of hazards: rotten timbers, rock falls, abandoned explosives, gas, and nests of venomous creatures, not to mention winzes full of water but covered with a film of floating rock dust which deceptively resembles a concrete slab.

whose vertical legs are the shaft and its shaft collar, the latter topped by a substantial platform equipped with tracks or a sheet-iron floor. The hoisting engine itself is in an ell or small shanty some distance away, with the hoisting rope (a wire cable or braided belt) coming off the drum of the hoist, passing over the sheave (pronounced "shĭv"), a large wheel surmounting the headframe, and so down to the lowest depth (the sump) to be hoisted from (Fig. 39). In cold climates the whole

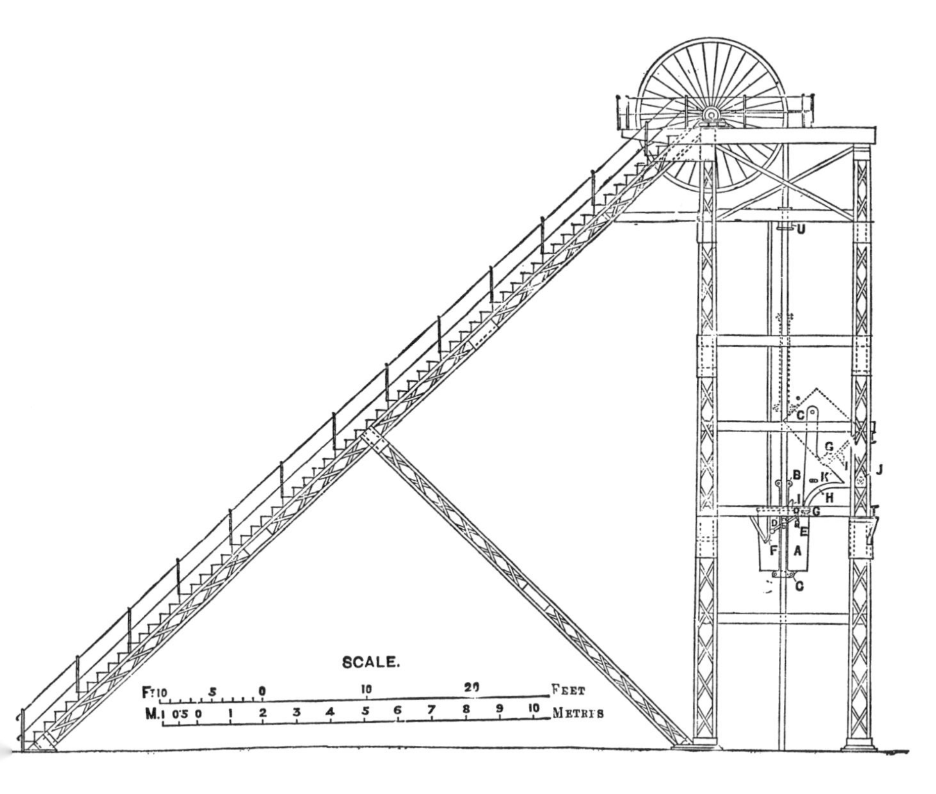

Fig. 39. A headframe and sheaves with self-dumping skip. From C. Le Neve Foster, *A Text-Book of Ore and Stone Mining* (1897).

edifice including the hoist room is sheathed for warmth. The shaft may be single or multiple. If the latter, one compartment or pump shaft is devoted to housing the pump rods, with as many cage compartments or manways as the size of the operation requires.

The visitor might ride a man car into a tunnel adit, or, in the initial stages of development, he might get into the sinking bucket to descend the shaft. This bucket abounds in mining camps; it is the size of an iron barrel, distinguished by a hoisting loop in the substantial bail and a shape like that of a truncated egg, the better to avoid being snagged by rocks projecting from the sides of the shaft. It is an all-purpose container designed to lower men and materials and is capable of hoisting waste, ore, and water. When the mine is in production, the bucket is replaced by a cage, a multidecked contrivance with stub tracks accommodating a mine car, and room for materials or a dozen men. Hoisting is done by a three-man team: the hoist operator, the cage tender, or cager, who rides the cage (Fig. 40), and the top lander. The cager deposits the down loads at the proper levels, then loads on cars of ore and waste, sending them topside to be removed by the top lander, who replaces them with empties or supplies. In mines handling considerable tonnage, the shaft may be sunk at an angle with the vertical to form an incline up which ore is hoisted in a wheeled skip, or self-dumping ore container, which is a cross between car and bucket. By and large, small mines have tunnel adits, since these are self-draining and require no expensive hoisting equipment, but well-financed mines lean toward shafts or inclines up which steam hoists can efficiently raise large tonnages of ore and waste. The foot of the shaft invariably houses the sump, or lowest point in the mine, where ground water naturally collects and where the internal pumping machinery (if needed) is located in convenient proximity to the engine house on the surface.

Adits and shafts are primarily means of entrance, access, ventilation, and transportation, rather than mining as such. They are driven without great regard for considerations other than economic necessity and cost, so tend to be laid out in straight lines or along proper tangents. In addition, crosscuts are added as the situation warrants. These are short, lateral tunnels which interconnect major workings, and are driven for exploratory, communication, and ventilation purposes. They are very often equipped with wind doors or air doors, which may be opened and

Fig. 40. The cage tenders. Reproduced by permission of Buck O'Donnell and Shaft and Development Machines, Inc.

shut to control the ventilation draft. The final work in this category is the winze, or internal shaft, likewise employed for communication, hoisting, or ventilation, but now and then used to follow a "hot lead" discontinuity, or fault which is filled with mineralization.

The real business of the mine is conducted in the drifts, or working tunnels, which follow the lode as nature and the mine superintendent dictate. One might, therefore, conclude that the lower the proportion of access workings to drifts, the greater the efficiency of the mine, but this conclusion is not necessarily correct. The Spanish rat-hole miners did almost nothing *but* drift or sink on ore, their workings winding in and out in labyrinthine fashion; so to this day do small-time operators. Engineers of a well-managed mine, however, attempted to work out the shape and size of its ore body so as to plan and construct access workings to the best over-all advantage.

Theory and reason combined to demonstrate that however time-consuming and arduous might be overhead drilling, the cost and trouble was well worth it if directed toward the end of moving as much ore and waste as possible by gravity. Accordingly, a well-organized mine might almost be said to have been worked from the bottom to the top, with its levels interconnected by winzes, chutes, and transfer raises into which the ore could be dumped so as to emerge at the lowest levels for direct loading by the resident chute tapper into the mine cars or skip (Fig. 41). The upper ends of these winzes and chutes were usually covered with a heavy grill of timber, through which the muckers dumped their Irish buggies of ore and waste that had been mucked into

Fig. 41. A chute tapper filling a skip. From C. Le Neve Foster, *A Text-Book of Ore and Stone Mining* (1897).

them in the working chamber. Unfortunately, shafts could not be so guarded, and they represented a serious hazard—to experienced miners. It was notorious that many a veteran miner had pushed a loaded ore car into an empty shaft, and followed it downward with his hands still firmly grasping the rim. Such accidents never befell novices; they were only too well aware of where they were and what they were doing, but old-timers tended to become absent-minded and begin thinking about other matters than whether the hoist cage actually was at their level or whether they merely supposed it to be.

Now and again a lode dramatically enlarges to form a pocket of considerable size. Both during and after mining out, such a pocket is called a stope; the process of eviscerating it is termed stoping out. The miner may approach it from above, drill and shoot downward, and elevate the ore to muck it out. This process is called underhand stoping, and is considered less desirable, because of the hoisting involved, than approaching from below, drilling up, blasting down, and mucking the ore on a common level with the extraction drift (termed overhand stoping). Supporting the roof or back of a stope can constitute quite a problem, and in the old days it was most often resolved by leaving pillars of low-grade ore or country rock at judicious intervals—an adaptation of the room-and-pillar system of coal or salt mining. On occasion the pillars might be constructed of rough masonry or log cribbing, piled up log-cabin fashion (Fig. 42). Almost invariably when the mine was worked out, the last act of the management or the first act of a scavenger would be to shoot down the ore pillars for their mineral content. Once that was done, the back of the stope would eventually collapse, possibly leaving a craterlike glory hole at the surface of the ground.

The term glory hole does not mean, as many believe, a very rich deposit. Strictly speaking, a glory hole is merely a conical pit of substantial dimensions whose sides, being unsupported in any way, tend to slump into their natural angle of repose. One can be created from below unintentionally, as by the collapse of a stope, or deliberately, by a controlled but analogous caving process. On the other hand, a mineralized outcrop which has been mined for a time by open-cast methods is said to have been glory-holed. The Silver King, near Superior, Arizona; the Gregory Lode, near Central City, Colorado; and the Yellow Aster group, at Randsburg, California, afford examples of glory holing.

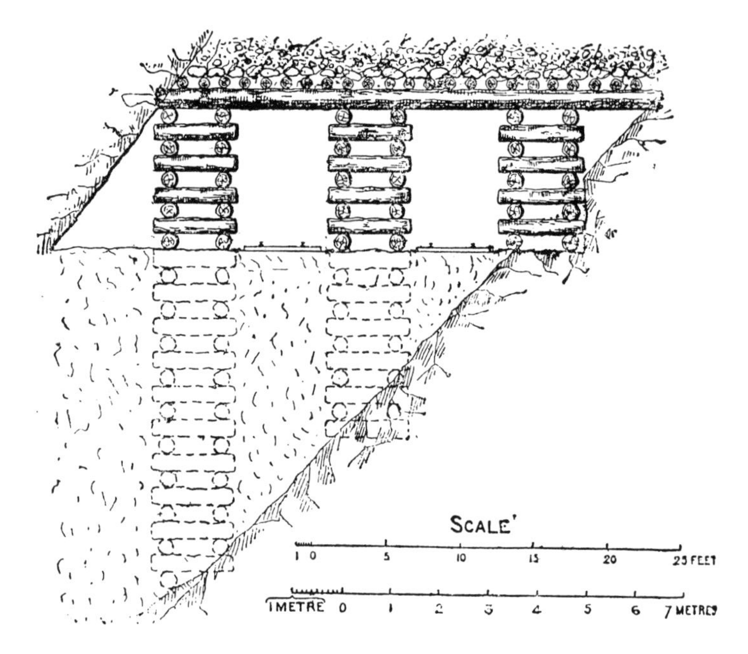

Fig. 42. Cribbing. From C. Le Neve Foster, *A Text-Book of Ore and Stone Mining* (1897).

Open-cast work must cease as the sides of the pit approach each other, whereupon the next step is to sink on the ore body by means of a shaft or incline moved a convenient distance from the outcrop. The ore body is then worked by conventional drift mining. More often than not the strike of a substantial lode can be made out upon the surface by aligning the outcrop with the headframes of subsequent works, which are almost always on a straight line with each other.

When stoping under average conditions of ground, the hoist shaft was first sunk to a convenient depth through or beside the ore body. From the shaft a level heading was run under the entire length of the deposit. This heading not only disclosed the extent of the ore body in one dimension but would later be improved to serve as a haulage and drain way. The stope was next extended laterally from the heading to the width of the lode, a task for which drilling patterns were modified considerably. Instead of double-jacking, a long row of miners single-jacked side by side, putting in edger holes along the top and bottom sides of the heading (Fig. 43). As the blocks of ore were shot out sideways, a heavy series of eight-foot cubical timber sets were installed to support the low but widening chamber (see Plate 6).

When the length and width of the ore body had been developed, it was time to begin overhand work. A chamber was driven as a raise upward at some convenient point and then timbered and floored with heavy planks to serve as a working platform. From this platform each eight-foot cube of ore adjacent was drilled and shot, to fall into the space below. In the vacancy so created, another timber set was placed, another platform installed, and the next block of ore prepared for removal. From time to time manways and ore chutes could be arranged, leading down to the haulage way below. To help support the back of the stope, strategically located sets were filled with waste, which was held within the timber cubes with heavy lagging. Surprisingly enough, these columns of waste-filled sets were actually stronger than the pillars of ma-

Fig. 43. Single-jacking. Reproduced by permission of Buck O'Donnell and Shaft and Development Machines, Inc.

sonry favored by Mexican miners for the same purpose. Though they would settle slightly as the ground settled, they would soon stabilize and support the back firmly. Where feasible and economical, when its active life was over, the entire stope could be filled with gravity-dumped waste, an effective measure in minimizing surficial slumps or ground movement affecting active portions of the mine (Fig. 44).

It is well at this point to distinguish between the often-confused terms waste and tailing. Heaped outside the shaft collar or portal is the dump of waste, composed of broken country rock which has been removed in the course of opening the mine and in drifting. Access works produce nothing but waste; drifts produce a mixture of waste and ore. In hand-drilling days, waste and ore were sorted on the spot at the face and each loaded in its appropriate car. In wet mines, in drifts which skirt the ore body, or in low-grade operations where the eye alone will not serve to tell one from the other, the distinction between ore and waste is made by means of a constant series of assays. The shifter tells the carloader (mucker or chute tapper) the results of the assays, and the loader then stands a stick in the middle of each ore car as a signal to the top lander to make appropriate disposition: waste to dump, and ore to ore bin. This preliminary sorting keeps to a minimum the amount of worthless tonnage sent to the mill, and development waste can often be distinguished by its rough, sizable chunks, dumped just as it was loaded after being mucked out of the tunnel or shaft.

Tailing, on the other hand, consists of finely ground ore particles from which the mineral content has been extracted. The tailing heap constitutes the chief ornament of the concentrator or mill and often enough appears to be merely a slope of dead-white silt, although it is occasionally stained in a gaudy fashion by chemical action on the metal residues. Mineralogists and gemmologists are fond of haunting waste dumps, seeking what they can find; but tailing heaps are worthless for their purposes, the contents having been ground to sand or flour. On the other hand, tailing almost always contains a proportion of mineral which was impossible or uneconomical to recover at the time but which, should extraction methods or market demand improve (and they usually do at some time), may one day experience a new lease on life as it is run through the machinery once again.

On the old mineral frontier ore and waste were almost always re-

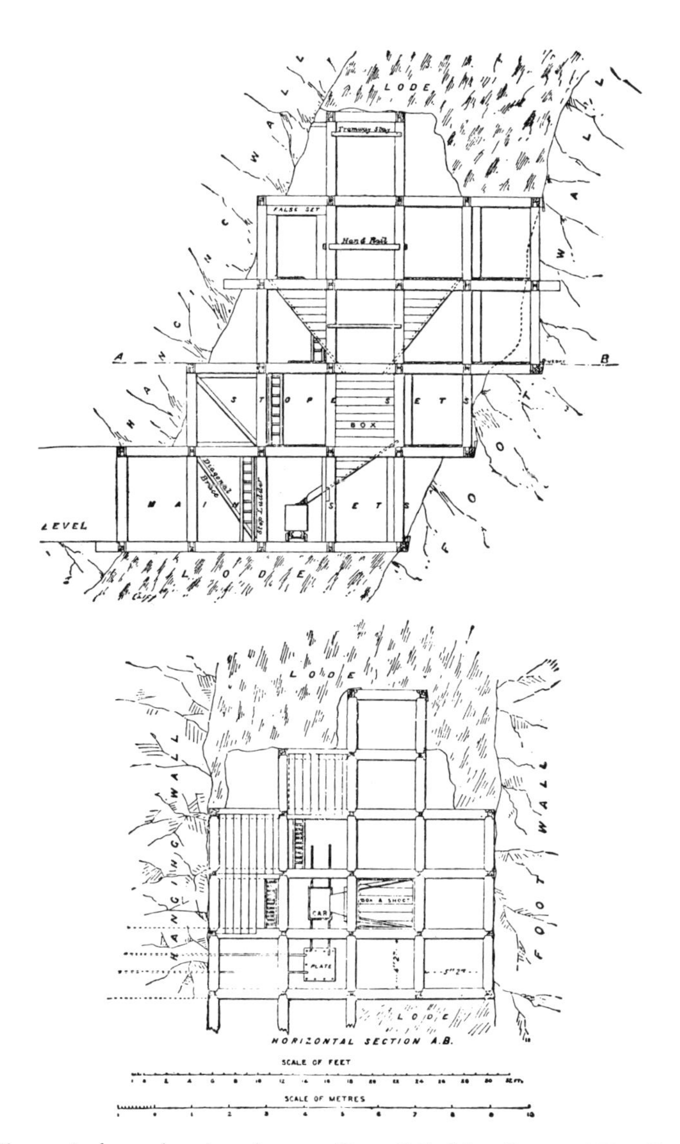

Fig. 44. A plan and section of a stope. From C. Le Neve Foster, *A Text-Book of Ore and Stone Mining* (1897).

moved in small, one-ton ore cars, specimens of which can be found almost anywhere in the West.[7] They could be used singly or linked by

[7] The common mine car dumps endwise when its hand-operated latch is released, but it also swivels on a central (vertical) pintle and can be dumped side-

short chains and hooks at each corner, and their four, closely set flanged wheels ran on miniature rails of about eighteen-inch gauge. The equipment was standardized, portable, as easily assembled as a child's model railroad set and but little more expensive, and was powered horizontally by man or mule.

The mine mule was an institution in himself. The operators favored small Spanish mules, much smaller than the Missouri giants employed by freighters. The British stannaries and collieries used ponies, but mules were far more intelligent and far less given to hysterics in a bad situation. A mule, moreover, would duck its head when its long ears encountered an overhead obstacle, whereas a pony would toss and beat out its inconsiderable brains. A mule, to be sure, was infinitely more self-willed, nor were its rear hooves to be despised, but the animal was capable of detecting trouble and giving warning to the miner wise to its ways. Some miners grew fond of their mules, petting and cosseting them familiarly, but most of the livestock was regarded with the same feelings as were nourished toward visitors: inescapable and inevitable nuisances (Fig. 45).

Mules were used at Bisbee as late as 1931; they pulled five cars, and their drivers would put the cars together tightly so that each would have slack with which to start them. They could count (the mules, not the men), so when the jolt of the fifth car was felt, they were off. If a sixth jolt came, they would balk. To fool them, the driver would spread the slack out of the last chains with his hands. Since the mules could not tell the difference between the load of one more ton and two more (with only one jolt), the result was no balk. A traditional miner's drinking song celebrated the relationship (Fig. 46) in the words:

> My sweetheart's a mule in the mine
> I drive her all day without lines
>   On the car front I sit
>   And tobacco I spit
> All over my sweetheart's behind.

Nonetheless, mechanized hauling moved the most tonnage for the

ways alongside the track. The bigger mines of the Comstock Lode used side-dumping cars of two-ton or greater capacity, of which they were inordinately proud.

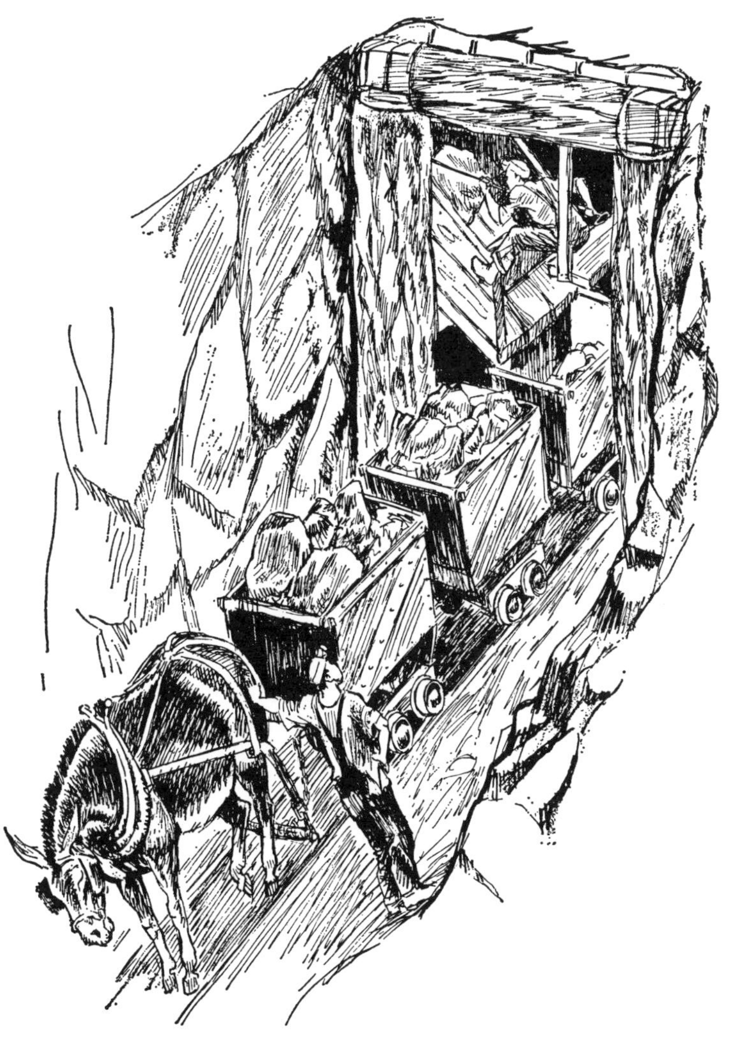

Fig. 45. The ore train from the stopes and drifts. Reproduced by permission of Buck O'Donnell and Shaft and Development Machines, Inc.

Fig. 46. The mule skinner. Reproduced by permission of Buck O'Donnell and Shaft and Development Machines, Inc.

lowest unit cost. The headframe with its giant sheaves is the most prominent feature in almost any representation of the ordinary mine of the frontier period. The boss of the ranch was the hoist operator, a man of unusual adroitness and sobriety who considered himself a notch or two above the common run of mine employees. At first he was signaled by a wire-actuated gong to which he could make no reply other than to jiggle the cage to indicate "understood." Printed broadsheets with the gong code were posted at every level station and at the shaft-collar and hoist-engine shanty. This rough-and-ready system was later improved by the installation of two-way electric annunciators.

Bell signals were fairly uniform from the first, being one of the traditional mining institutions imported without modification from Cornwall. Time and experience indicated that such signals ought to be standardized by law and mining code not only to prevent misunderstanding by new hands but also to enable hastily brought in rescue teams to function without slip-ups. A portion of the Arizona State Code of Mine Signals follows:

ARIZONA STATE CODE OF MINE SIGNALS

1	Bell	Stop immediately if in motion.
1	Bell	Hoist muck.
1	Bell	Release cage, skip or bucket.

2	Bells	Lower.
3–1	Bells	Hoist men, if bells rung slowly.
3–2	Bells	Lower men, move slowly.
4	Bells	Steam on or off.
5	Bells	Blasting or ready to shoot. This is a caution signal and if the engineer is prepared to accept it he must acknowledge by raising bucket or cage a few feet then lowering it again. After accepting this signal engineer must be prepared to hoist men away from blast as soon as the signal "1 Bell" is given and must accept no other signal in the meantime.
6	Bells	Air on or off.
7	Bells	Danger signal followed by station signal calls cage to that station. This signal takes precedence over all others except an accepted blasting signal.

STATION SIGNALS

1–2 Bells, Collar of Shaft	2–1 Bells, 4th Level
1–3 Bells, 1st Level	2–2 Bells, 5th Level
1–4 Bells, 2nd Level	2–3 Bells, 6th Level
1–5 Bells, 3rd Level	2–4 Bells, 7th Level . . .

Station signal must be given before hoisting or lowering signal. The engineer shall not move a cage, skip or bucket unless he understands the signal. One copy of this signal code shall be posted on the gallows frame, one at each station, & one before the engineer. Special signals may be used provided they are easily distinguished by their sound or otherwise, from the foregoing code, and do not interfere with it in any way.[8]

Not given in the code, perhaps because it was almost sacredly traditional, was the bell signal 3–3–3. The miner's dirge, it was given by rescue teams and invariably meant "Hoist recovered body."

The hoist operator's companion was the top lander, so called because he worked "on the grass" in the mine yard. His duties consisted of removing cars of ore and waste from the cage, pushing each to its appropriate destination, and loading such materials as the cager or a given crew's tool nipper might require. Like the hoist operator, he was most often a miner who had been crippled too seriously to work underground, but not so seriously that he had to be given a sinecure, such as the job of watchman.

[8] From a placard in the Iron King Mine, Humboldt, Arizona.

The problem of what to do with tailing was the business of the millman, but waste disposal was always a question for the mining engineer. In the early days waste was a dead loss from first to last. A certain amount could be used for surfacing access roads, as fill for stope cribbing, and for other minor purposes, but most had to be hoisted and dumped, and the engineer whose dump assumed expansive proportions too swiftly had much to answer for at the next board meeting. (Today miners sensibly put as much waste as possible right back underground where it came from, as fill. On the other hand, today they employ all sorts of earth-moving machinery that was unheard-of on the mining frontier.) Happy the syndicate whose adit was high on the face of a slope, as at Tombstone or the Comstock. Waste could simply be pushed out the adit, run along the rails which overlaid the dump, then spilled sideways.

In flat land the disposal problem was presently settled by building the shaft platform about ten or twenty feet above the mean surface level. Development waste from the initial shaft sinking was used to build a viaduct off to one side of the headframe; at its end was installed an upgrading shanty and ore bins over the haulage road (Fig. 47). Ore could be dumped by gravity directly into the wagons, which took it to the mill, and cobbed-off waste was tossed out the far door. As soon as this work was completed, another dump was commenced in another direction, to be continued until the mine ran out of space or of ore.

The problems of pumping and timbering almost always arose as soon as ground had been broken. While in the West water was seldom the obstacle that it was in the eastern collieries, it could be bad enough; at the Comstock it nearly proved fatal. As a rule it got into a mine by percolating down from rain or surface accumulations to find and fill a reservoir of permeable strata—loose sandstone or some such blotterlike and porous material. If recent igneous activity had occurred in the vicinity, it might heat the water and charge it with dissolved minerals, thereby increasing its penetrative ability. This subsurface water then hunted out and trickled into any fault or discontinuity it could find, even dissolving the natural cement which held different strata together. Sooner or later a substantial flow of water began to enter the mine. At that juncture pumping was definitely indicated.

As has been noted, the *sistema del rato* was helpless to do much about any substantial amounts of water other than hoisting or bailing. By 1850, however, steam pumping had been established in western Europe for nearly a century. For that matter, steam engines had been developed to pump mines and only subsequently were adapted to moving heavy

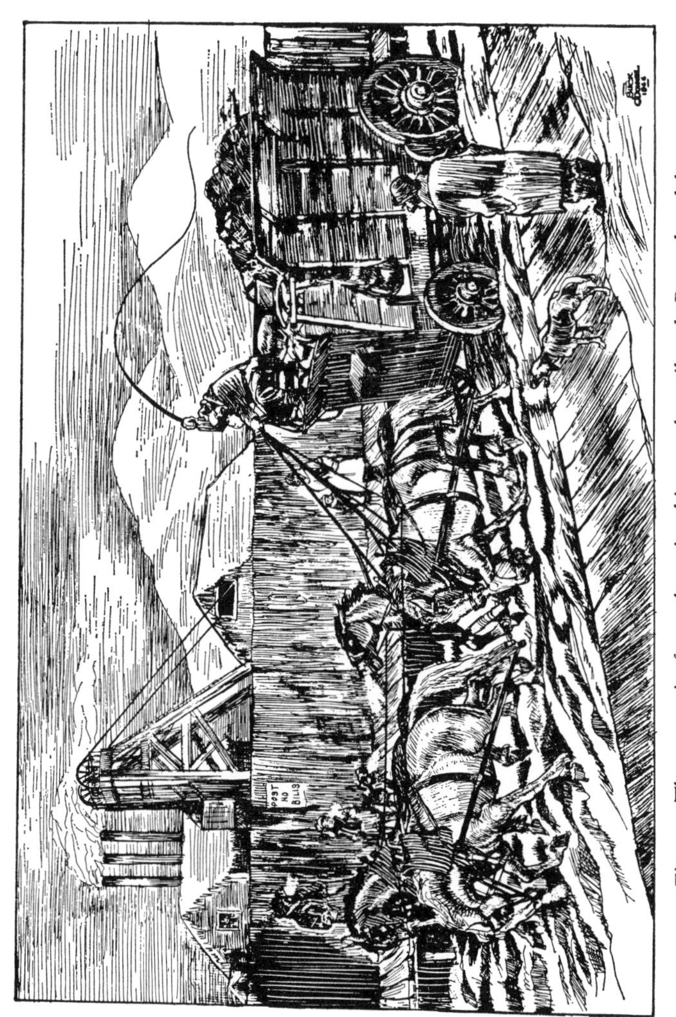

Fig. 47. The ore train from the mine bins to the railroad. Reproduced by permission of Buck O'Donnell and Shaft and Development Machines, Inc.

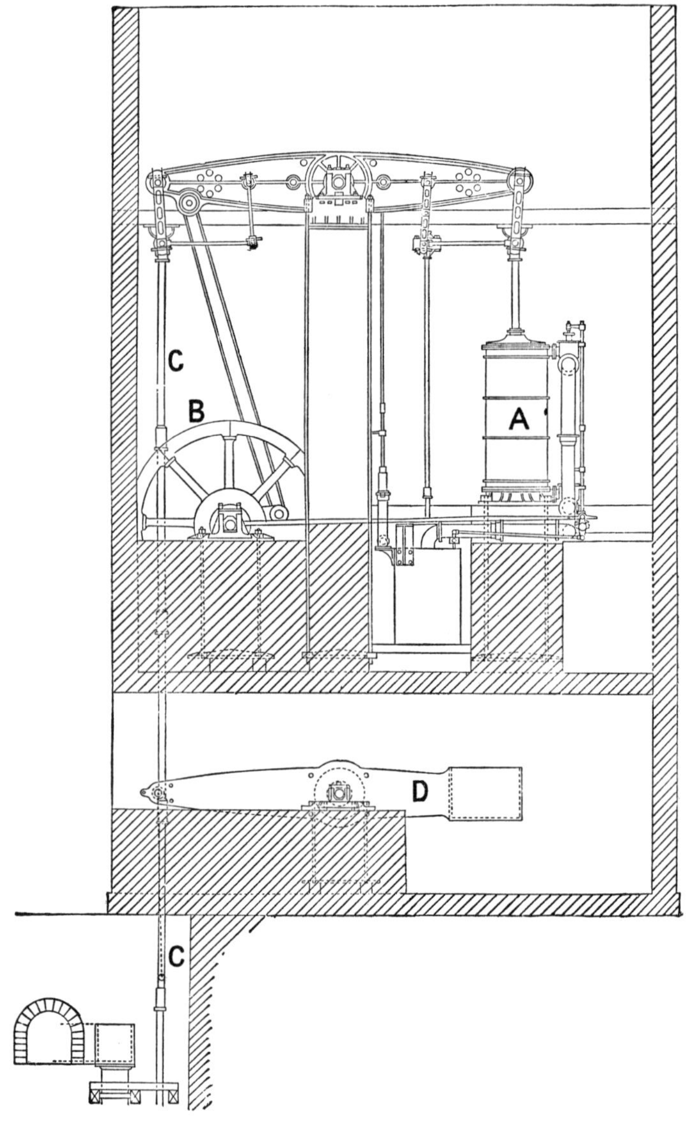

Fig. 48. Cornish beam engine with flywheel. *A*, steam cylinder; *B*, flywheel; *C*, pump rod; *D*, counterweight. From C. Le Neve Foster, *A Text-Book of Ore and Stone Mining* (1897).

objects about. The chief dewaterer of the mineral frontier was the Cornish pump, whose devotees and detractors alike approached near fanaticism in their judgments upon the system. A typical pumping setup was erected upon a remarkably massive foundation next to, and extending a goodly distance down, the pump compartment of the shaft. It was usually powered with a vertical-cylinder walking-beam engine which drove a flywheel and occasionally high-ratio reduction gearing. The large bull gear was in turn connected by a pitman rod and bell crank (or bob) to the pump rod running vertically down the pump compartment all the way to the sump at the lowest level of the mine (Fig. 48).

The pump rod was both the heart and the Achilles' heel of the Cornish system. It was made of the very best grade Oregon pine, sometimes as much as sixteen by sixteen inches in cross section, spliced, bolted, and reinforced with all manner of iron fittings. Massive but variable counterweights were pivoted to the rod and hung on the free ends of the bobs at suitable points in order to secure the desired degree of static balance in the system. The rod was held in alignment in the pump compartment by timber guides spaced down at frequent intervals. Large balks of timber were fastened crosswise here and there to the rod so that if—

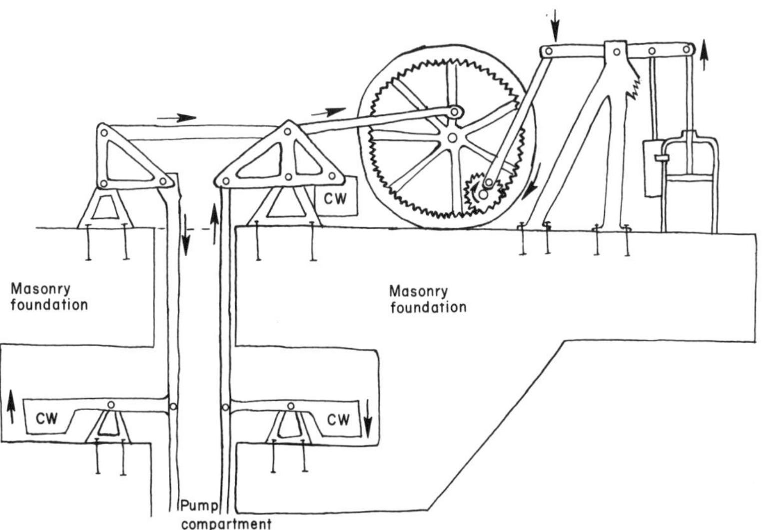

Fig. 49. Schematic diagram of rotative Cornish beam engine with bull-gear flywheel and twin pump rods with bobs and counterweights.

or when—the rod broke these balks would jam in the compartment timbering and so prevent the entire rod from collapsing like a handful of jackstraws into the sump. Finally, by ingenious arrangement of strategically placed bobs, twin rods could be worked, run down inclines, or even be made to turn corners (Fig. 49).

In the very deep mines the rod sprouted stages of pumps down its length, each pump raising the water from one successive pumping-station level to a reservoir and pump on the next. The lowest pump, located in the sump, raised the water by direct lift atop its piston; the rest of the pumps, farther up the line, were equipped with plungers which pushed, rather than raised, the water (Figs. 50 and 51). Being

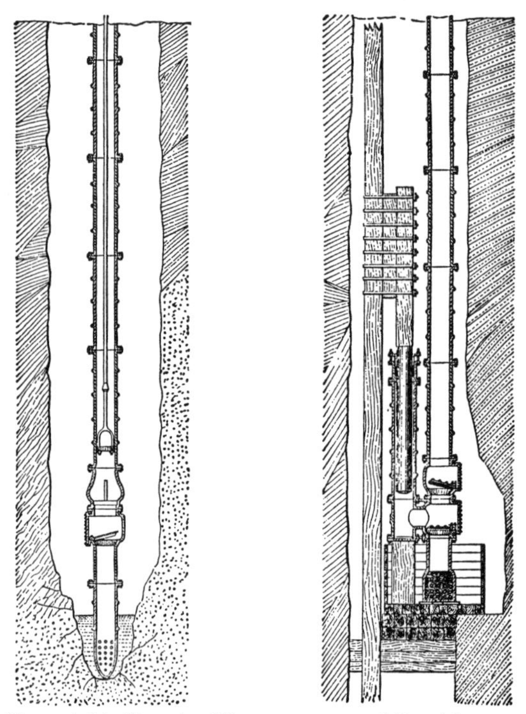

Fig. 50. (Above, left). Drawing-lift sump pump of Cornish pump pitwork. From C. Le Neve Foster, *A Text-Book of Ore and Stone Mining* (1897). Fig. 51. (Above, right). Intermediate-stage pump of Cornish pitwork. From C. Le Neve Foster, *A Text-Book of Ore and Stone Mining* (1897).

submersible, the Cornish pump worked fairly well, when it worked, but its capacity was decidedly limited. The standard version completed six or seven cycles per minute, each downstroke raising about one net ton of water. Since the fantastic mass of the inherently weak rod prohibited any pace that was less deliberate, increased capacity could be gained only by enlarging the steam cylinder or by adding additional cylinders: perhaps the largest Cornish engine ever built was a one-hundred-inch giant in the Comstock Lode. A peculiar feature of the system was that the engine worked only to raise the rod and water in the sump level. On the downstroke the weight of the assembly (often dozens of tons) did the effective work of pushing up the water.[9]

Over the years the Cornish pump became ever more complicated but no whit more reliable. Ingenious uses were made of it. By attaching steps to the rods (when used in pairs) or steps to one rod and fixed platforms to the compartment timbering, a dexterous man could "ride" up the rods elevatorwise (Fig. 52). On the other hand, let anything choke the pump intake or cause it to suck air, and the next thing the operating engineer knew the rod would snap and let the whole system race free of load, often concluding with a Grand Moral and Educational Explosion of the Flywheel.[10] Indeed, pump breakdowns from necrotic pitwork were one of the most frequent causes of technical trouble in the mines, and it is well to note that the claim that Cornish pumps were uncommonly economical to operate (nine cents per ton of water raised) was usually founded only upon analysis of fuel consumption while disingenuously omitting such matters as capitalization and installation, maintenance, and general losses to the operation occasioned by down time for frequent overhaul and major repair.

It was not lost upon visitors to the mining camps that millions of gallons of mine water were being expensively pumped up and then let go down some arroyo, while at the same time yet more huge quantities of water for man and mill were being imported at enormous expense. The assumption was often made that what was being pumped up should be just the thing for communities and concentrators. So it was at, say,

[9] Perhaps the best brief description of the Cornish pump can be found in Elizabeth L. Egenhoff, "The Cornish Pump," California Division of Mines and Geology, *Mineral Information Service*, Vol. XX, Nos. 6, 8 (June, August, 1967), 59–70, 91–97.

[10] Grant H. Smith, *The History of the Comstock Lode, 1850–1920*, University of Nevada *Bulletin*, Vol. XXXVII, No. 3 (July 1, 1943), 245.

Fig. 52. Miner riding up pump rods. From Louis Simonin, *La vie souterrane* (1867).

Globe, Arizona. There are those who claim that Tombstone water is drinkable (especially if tinctured with sufficient quantities of corn liquor). Unfortunately, most mine water was so charged with muck, arsenic and sulphides as to be unfit for man, beast, or boiler—or, to judge from photographs of some of the crews, even personal ablutions.

The question of support and timbering had to be settled on the spot in light of geological and economic conditions. The Colorado mines, which were drifted into comparatively monolithic granite, needed little support, since the heading itself in such ground forms a natural arch. On the other hand, the Comstock Lode required the most novel and gigantic timbering ever devised. The condition of the average mine lay between these two extremes, yet it can be said that the western mines were certainly better off in this respect than were the horizontal, inclined, "blocky,"[11] and highly stratified sedimentary layers typical of coal operations. This was just as well, for appropriate timber was cheap and plentiful in the East where it was most needed, but often quite scarce in the usual western mining district. Costs being what they were, the small western operator was prone to economize where he thought he could get away with it and heavily reinforce only areas of obvious weakness.

The average drift's tunnel-set timbering would have evoked no surprise from Herodotus. The timbers averaged eight feet in length and about eight by eight inches in squared cross section. At the time of installation they were cut to size with a framing saw; then the "cap" or top timber was shallowly flattened (dapped), with a special tool which resembled a pick crossed with an adz, toward each end to catch the heads of the side pillars or posts. The posts were raised and blocked into place with wooden wedges or rock waste, then the cap set atop them. Fortunately, recently opened ground does not normally take weight immediately, so the cap need not be a tight fit. When the ground did settle, the cap would then be firmly wedged into place. Care had to be taken, however, that the posts were set truly vertical and that any tendency for them to wobble was eliminated. The sets were spaced about five feet from center to center, with collar braces or spacers run between the corners of the sets to ensure rigidity and to take the end-on thrust of blasting as the heading was advanced. Finally, lagging (two-inch plank roof and walls) was installed to box in the drift.

Now and again tunnel sets were supplanted by stulls, which were roughly equivalent to free-standing stanchions. These heavy props were not interconnected but were erected at right angles to the hade (vertical

[11] Large blocks of the hanging wall are prone to detach themselves in this sort of ground, particularly after ground water or air slaking has weakened the bedding cement between strata.

Fig. 53. Timber staff. Reproduced by permission of Buck O'Donnell and Shaft and Development Machines, Inc.

dip) of the formation—that is, at nearly ninety degrees to the plane of the hanging- and footwall of the lode. The trick of installing stulls was to set them so that ground movement would tighten them up, bringing their stance closer to the desired ninety degrees rather than otherwise, so as to impede the same ground movement which was locking them tighter. They could be used to support working platforms and sometimes to brace up a heavy timber umbrella which supported a patch of loose ground in the back of the stope. Like sets, stulls were tailored on the spot with the assistance of a gauge made of two long wooden battens sliding telescope-fashion through a flat metal clamp. The miner extended the battens across the distance to be timbered, tightened the clamp, and then measured for his saw cut directly from gauge to timber (Fig. 53). Mine owners purchased these sets and stulls from mountain sawmills by the wagonload, and in desert country the price per thousand was a substantial part of the operational costs.

Adequate ventilation was always a matter of deep concern to those who delved beneath the earth. In all cases the object was to ensure that as great a volume of fresh air as possible circulated through every corner of the mine in as short a time as feasible. This goal was fairly easy to achieve in those workings so laid out that the ventilation air could be forced in at one extremity and drawn out at the other, assisted by the judicious placement of crosscuts and wind doors. Dead-ended drifts and headings had the highest concentration of powder smoke, human exhalations, and rock dust to be found in the mine; yet not only was the air in those areas naturally stagnant but complete circulation had to be achieved in the least time—an hour at most, between shifts. In hot or wet mines the problem was compounded, but at least western developments were usually free of the white damp, the explosive methane gassiness characteristic of coal workings (Fig. 54).

During the initial development of a mine a fire could be maintained at the foot of the shaft, as was the practice in Neolithic flint mining, the ascending column of hot air creating a chimneylike draft. Open fires were not favored in mines as a rule, however (though underground blacksmith shops sheeted with boiler iron were occasionally found), and as soon as a separate ventilation inlet portal could be provided, steam-driven Root (lobe) blowers would be externally installed to afford low-pressure ventilation. The hoist shaft usually served as the outlet stack, since it was the longest continuous raise in the mine and hence possessed the greatest uplifting stack draft; visitors commented freely upon the heavy white column of vapor rising from its collar on cold days. Through this cloud the cages and crews could be dimly glimpsed going about their duties.

Before the advent of air drilling it was necessary in tunneling and

Fig. 54. The hot box. Reproduced by permission of Buck O'Donnell and Shaft and Development Machines, Inc.

mining to lay a high-pressure ventilation trunk into the heading or working chamber with its outlet placed as near the face as practicable. In hot headings the air could be appreciably cooled both by expansion and by the use of water sprays placed in the line. Air drilling in most cases rendered such trunking unnecessary, since the air machines supplied ventilation from their exhausts, which were likewise cooled by expansion. Even so, it was found desirable from time to time to install small blowers deep within the mine itself, although the record fails to explain how these blowers were powered.

In the huge Comstock mines most ventilation was the product of the natural circulation created by the stack draft of shafts over a thousand feet deep. In the first few years of development air entered the mines through the hillside adits, whose portals were about on F Street, and then was swept up the shafts in easy fashion to their headworks near D Street. When the levels went below this point, the workings were interconnected; adjacent mines whose operators might in all other respects be mortal enemies nevertheless found it desirable to lay aside their quarrels to deal co-operatively with problems of ventilation and pumping. Experienced Comstock miners found it easy to predict which shaft would carry the downdraft and which the updraft when the air doors were opened. Unfortunately, mine fires benefited by the same procedures. All too often the white cloud above the shaft collar would be replaced by wisps of gray turning rapidly into a black tower of smoke. As the hoisting works' whistles began to shriek and agonized families to gather in the mill yard, the black tower with a dull explosion would turn into a roaring column of flame hundreds of feet high. Often enough there would be no alternative to sealing the shaft and injecting live steam to smother the flames—and any unfortunate miner still alive in the workings. Peculiarly enough, these fires seemed to reverse the normal order of things, since, after the mine was reopened, the ventilation was usually found to proceed in a backward manner, with the former updraft shaft now carrying the downdraft.

CHAPTER VII

The Western Miner and His Methods

THE WESTERN MINER was something of a tramp. It was said that most mines needed eight men to handle the normal two-man task underground, since there were "two a-coming and two a-going, two a-rustling and two a-working." The tramp miner was a good hand when he took it into his head to be one. When he ran into a serious situation that was entirely new to the settled "home guard," he had only to recall what he had seen in Butte or the Comstock or the American Nettie, and with a workable basic plan the problem was virtually solved. But when payday came along, he might decide to leave at once by side-door Pullman to investigate the wholly unfounded rumor of "four dollars a day for a machineman" just over the range. It was wholly unnecessary to have a reason for leaving, but he might rise to the occasion by complaining that there had been a lump in the oatmeal that morning or that "the boardinghouse hotcakes are out-of-round."

It was just such individualists who made up the face crews that assembled at the headworks of a typical mine as the shift began. Mining costume was no model either of safety or of elegance.

Dress and Equipment

On his head he wore a battered felt hard hat of moderately wide brim, stiffened with resin until it approximated the modern construction helmet in its capacity to deflect light blows. He wore a short-sleeved undershirt and union drawers, a drab shirt of heavy weave, baggy trousers without visible means of support, and either boots or ankle-high brogans. If he elected to wear brogans, his trouser cuffs were usually tied below the knee with lengths of Bickford fuse. When sinking a shaft or otherwise laboring under wet conditions, he might wear a sou'wester and oilskins (Fig. 55). In the Comstock Lode the combination of high humidity and extreme heat dictated a costume of boots, hat, and a sort of breechcloth as the only concession to modesty. Before the day of

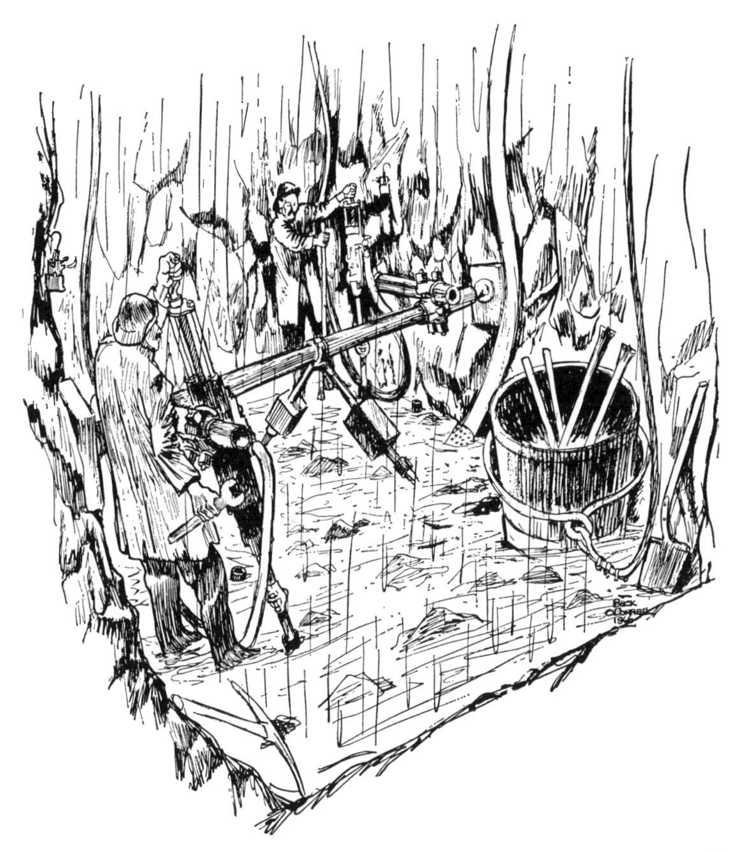

Fig. 55. The shaft sinkers. Reproduced by permission of Buck O'Donnell and Shaft and Development Machines, Inc.

change rooms the miner kept a fleece mackinaw on a nail in the level station to afford a measure of protection from pneumonia (a deadly killer at high altitudes) as he trudged his soggy way home through the cold and incessant winds characteristic of mining camps.

At shift end the miner smelled like one of the mine mules, from sweat, dust, and drip. Nevertheless, the miner, like most technical men, was quite tidy. Dirt-filled candle boxes, supervised by the tool nipper, were distributed to the stations to serve the needs of nature. Refuse of

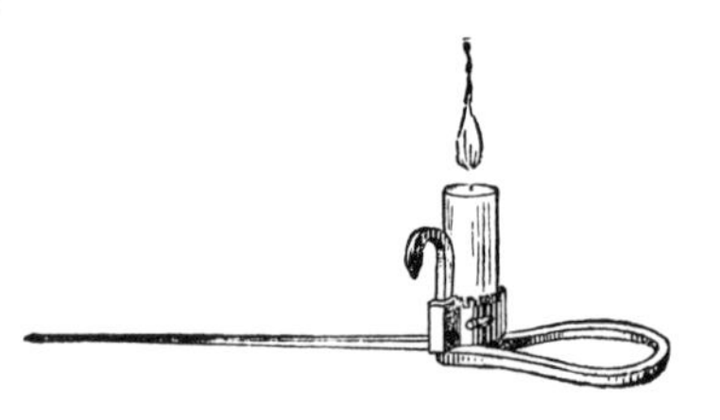

Fig. 56. The miner's candle holder. From C. Le Neve Foster, *A Text-Book of Ore and Stone Mining* (1897).

other sorts was carefully policed and disposed of. Miners spent much time finding and removing rats which had experienced *Todt und Aufklärung* behind the lagging. Working as he did in semidarkness, deafened by the clangor of drilling, the miner relied heavily on his nose for warning of danger, and detested extraneous odors that interfered with his ability to smell the first whiff of dreaded wood smoke that meant fire. And, of course, he never rested until the bodies of men killed in accidents were recovered and hoisted to the "grass" for decent burial.

For light the miners often used the oil lamp of antiquity, whose reservoir held enough oil for ten hours' burning time, the traditional length of the shift.[1] With a cylindrical body and a long, somewhat conical upslanting spout for the wick and flame, the oil lamp could be fastened to the forehead. Manufactured of tin and fueled with kerosene, it persisted for long on the frontier, but was prone to drip and start fires upon the slightest provocation. (A variation of the oil lamp was depicted in an early Christian catacomb painting which shows a miner holding a poll pick in one hand, and the classical "gravy-boat" oil lamp equipped with a short chain terminating in a spike in his other hand. This spike could be driven into a shoring timber or rock crevice and the lamp hung safely out of the way.) For the most part, American miners were supplied with three candles, whose combined burning time was ten hours. Two candles were stored in the bootleg; the other was inserted in a glassed lantern for transportation down windy shafts and galleries.[2] Once at the face or station, the candle was removed from the lantern and inserted into a wrought-iron holder of characteristic shape, likewise equipped with a

[1] Thus a mine employed two shifts a day, an hour being allowed each way for travel to and from the portal and for the smoke and dust of blasting to clear.

[2] Rickard, *Man and Metals*, I, 382, 444, 491.

long spike on the horizontal axis and a pointed hook on the vertical axis for handy use (Fig. 56). If extra candles were obtainable, the miner would drive a series of long nails into a piece of board so as to form a sort of circular palisade. The extra candle went in the middle of the nails, and his lunch bucket was set atop it in the manner of a chafing dish. In this fashion he would have hot coffee or, if a Cornishman, tea, during his half-hour lunch break (Fig. 57; see also Plate 7).

Fig. 57. The miner's coffee warmer. Reproduced by permission of Buck O'Donnell and Shaft and Development Machines, Inc.

As cage or bucket deposited each crew at its predestined level, work began with the miners mucking out the ore and waste left by the blasting of the preceding shift. As the round-point shovels of the muckers began fanning, the shift boss marked the point for the new drill holes in the face, either by means of a few blows with his geologist's hammer or, in later days, by a round spot of soot deposited by the flame of a carbide lamp. Drilling began forthwith for the next round, or series of shots (Fig. 58). Before 1875 drilling was conducted by hand sledging of simple steels. These sledges were eight-pound double jacks if worked two-handed; four-pound single jacks if plied by the right hand alone. A lone miner single-jacked, holding and turning the steel with his left

Fig. 58. Round in, round out. Reproduced by permission of Buck O'Donnell and Shaft and Development Machines, Inc.

hand while smiting with his right. Two or three men constituted a double-jack team, one turning (or shaking) the steel while the others pounded away in disciplined, rhythmic succession (Fig. 59). The steels themselves were merely round or octagonal rods with a plain, slightly flaring chisel-bit tip. They came in sets whose tip flares grew progressively narrower as the length increased: the bull steel or starter was about a foot long, but its bit ran nearly $1\frac{1}{4}$ inches from corner to corner, while each subsequent "change" grew in length by 6 inches while its bit decreased by $\frac{1}{32}$ inch so that it would "follow" in the hole and not get wedged, or fitchered. A 3-foot length with a $\frac{3}{4}$-inch bit was about the maximum. Upon descending to work, the miners carried whole

Fig. 59. Double-jacking. Reproduced by permission of Buck O'Donnell and Shaft and Development Machines, Inc.

bundles of starters and changes, numerous enough to be frequently replaced as the bits dulled at the rate of one steel for every 6 inches drilled. At the end of the shift the bundles were carried up to that busy man the mine blacksmith for resharpening and retempering.

In frontier days the drill was made of Black Diamond round or octagonal bar stock, cut into the desired lengths. The bit end was heated cherry red and flattened by being hammered into a width a little greater than the prospective hole diameter. Then the slightly curved cutting edge was hammered lightly into the desired angle—blunt for starters, more acute for increasing depth in the changes—and the corners beaten in to the exact diameter of the bore hole. The edge was touched up with a file to uniformity of profile and angle, since any irregularities would mean rapid fracture of the bit. Tempering was described by Lakes:

> [The blacksmith] took a piece of steel, heated it to a cherry red, laid it on the anvil and pounded it lightly with his hammer all over, to toughen it by blows, occasionally dipping his hammer in the water to "water temper" [the bit]; this further toughens it, by partially cooling it. Now the bar was again put in the fire and heated to a cherry red, care being taken not to keep the bar too long in the fire, as this would tend to take its toughness out, or produce blisters. The bar was [next] plunged about an inch into the water, and then rubbed against a brick, to show the colors plainer. These passed from the point upward, gradually through [a light straw, passing through shades of yellow, brown, purple, blue and red]; to arrest it by suddenly cooling it at "straw" would make it too brittle for ordinary drills, except a "bull drill" [starter]. Now the "straw" turns into a copper hue, a good point to cool off for a *drill*. Now it passes into a blue, at this point it would be well to cool off for a *pick*. The edge of a drill [bit] is almost of secondary importance to the *sharpness of the projecting corners*; when these are gone, the drill is used up, and clogs in the hole. Some rocks like sandstone will, by reason of the quartz in them, wear off the corners very rapidly, others, like limestone or granite, less rapidly.[3]

The Cornish miners used a wonderfully archaic system of drilling known to have been employed by Pharaonic tunnelers and quarrymen and probably taken to Cornwall by the Phoenicians. The Egyptians, who had few tool materials but wood, stone, and copper, used a suspended wooden ram armed with a dolerite ball head to bash open a cut

[3] Lakes, *Prospecting*, 274–80.

hole in the center of the face (tests have shown that they could make respectable progress at the rate of about $1/5$ inch an hour in granite), preliminary to wedging and chipping the remainder of the face down radially into the cut hole. The Cornishmen got the same effect by drilling a set of converging holes 24 to 36 inches long, aligned to meet at the apex of a pyramid in the center of the face. When shot, the holes usually primed each other by mutual concussion, producing a conical reliever cavity no different in principle from that of the Egyptians. Into this cut hole the rest of the face was squeezed down by the peripheral reliever, edger, and lifter shots which came along presently. Where ventilation was good, the Cornish miner could blast his cut-hole shots just before lunchtime on the shift, so that the smoke and dust would clear while he was consuming his tea and pasty and peradventure doing a bit of high-grading.

To judge by his meager evidence, Mark Twain's hard-rock mining experience was a compendium of everything the expert miner did *not* do. Apparently he and his partners merely drove one short hole into the center of their face, shot it out, and then were discouraged when the blast pulled only "a bushel or so of hard, rebellious rock."[4] Such malpractice would only minimize the force of the shot, progressively narrow the face, and finally reduce the practitioner to single-jacking at the point of a stony funnel under impossible conditions. Then, too, Clemens was sinking a shaft, making it difficult to clean out the rock dust which soon filled the bottom of his down hole, cushioning the force of each blow. Experienced miners turned the steel slightly with each blow, replaced dull steels frequently, and, when they could, poured a small amount of water into the hole to make a puttylike mud of the dust and cuttings. When done right, this mud would adhere to the steel, so that it could be withdrawn and given a sharp rap against a nearby rock to clean it. Up holes, though hardest to drive, kept themselves clean; down and flat (horizontal) holes might be drilled dry and kept clean with a miner's spoon of copper; the men could also clean the holes by mopping them with a frayed wooden rod, to the moistened fibers of whose tip the rock dust would cling.[5]

[4] Clemens, *Roughing It*, 211–13.

[5] Some critics have questioned this assertion, but it is to be found in the strongly Cornish-oriented work of C. Le Neve Foster, *A Text-Book of Ore and Stone Mining*, 160.

The Cornish drillers by virtue of their vocation were professional athletes and took as much pride as John Henry in their ability to drive steel. For that reason, drilling contests became organized public festivities throughout the mining West, with each district sending its champions to the national eliminations to compete for substantial purses. The events were single-jacking and two-man double-jacking upon a block of stone which tradition insisted must be only Gunnison granite from Colorado. More than brute strength was involved in the contest to drill as deeply as possible in a fixed time—usually fifteen minutes. The double-jack teams practiced for months to perfect not only steel changing, but even to exchanging places without losing a lick. It scarcely requires mention that frequent refreshment and some wagering of money were prominent among *aficionados* of this manly art (Fig. 60).

Fig. 60. The drilling contest. Reproduced by permission of Buck O'Donnell and Shaft and Development Machines, Inc.

Blasting

Toward the end of the shift it was time to tally and shoot the face as the traditional final duty. In black-powder days, the blaster began by checking about carefully for abandoned tools or any other impediments to a dignified departure and arranging his paraphernalia to his satisfaction. He then loaded the drill holes with the paper cartridges which he had prepared aboveground or in a secluded station of the level, filling them from the powder keg in the midst of royal solitude. He always worked with wooden or copper tools, since iron-induced sparking was rightly considered unsafe. After his holes were loaded, he stemmed or tamped the charges with damp clay or drill cuttings and great discretion, for too much stemming reduced space available for the powder load and hence reduced the amount of rock the shot would pull, whereas too little stemming permitted the shot to blow away out the hole to even less effect (Fig. 61).

The blaster on the American mining frontier employed Bickford slow-match fuse for the critical and delicate business of igniting the round of shots. From the employment of gunpowder for the first time in blasting until 1831, ignition was a chancy and terribly dangerous problem. The military slow match of saltpeter-impregnated cord was not suited for blasting, since it burned entirely too slowly. Miners devised various metal tubes, hollow wooden rods, and even split reeds cleaned of their pith and bound back together with string to contain their igniting composition, all tending either to misfire or fire too quickly. Saxon miners even employed a grooved wooden cone to serve both as stemming and as a conduit for priming composition, but there is no suggestion that this device worked any better than the wooden priming tube, around which the stemming was pushed and poked. There was little wonder that miners, like military gunners, fervently invoked Saint Barbara each time a shot was fired.

In 1831, William Bickford, of Tuckingmill, Cornwall, invented the first reliable fuse, which has since borne his name.[6] It was manufactured

[6] Today there is also a Bickford-Cordeau product which looks and smells like conventional fuse but definitely is not. It is composed of a fuselike covering about a core of PETN, and is better known as Primacord-Bickford. It is not ignited but detonated, and carries detonation almost instantly from charge to charge. It is useful in all forms of engineering, military and civil, but now and then the newspapers carry stories (usually obituaries) of unwary amateur miners who have mistaken Primacord for the older product.

Fig. 61. Loading the round. Reproduced by permission of Buck O'Donnell and Shaft and Development Machines, Inc.

by twisting six strands of jute about a central core of powder, wrapping the whole with strong twine, and wrapping again with one or even two layers of waterproof tape. The resulting product was flexible in nature and in use and fairly reliable, and when it failed, it tended to fail safe rather than to induce premature blasts.[7] This fuse, carefully secured

[7] Edgar Taylor, "High Explosives and Safety-Fuse," *Mining and Scientific Press*, Vol. XCVIII (May 22, 1909), 726–27. See also Otis E Young, Jr., "Fire in the Hole; Evolution and Revolution on the Western Mining Frontier," *American West*, Vol. VII, No. 4 (July, 1970), 15–19.

to the head cartridge in the hole, was measured off its spool and cut, leaving a rattail of precise length hanging down the face from the collar of the hole. A good round was one which pulled the maximum amount of rock. That was best accomplished by ensuring that all shots of the same nature (that is, cut holes, relievers, edgers, and lifters) went off as nearly as possible at the same time; hence precise fuse cutting was of great moment. Shots followed a four-phase sequence, provided the cut hole had not been shot during the lunch hour. The cut hole had to go first, the relievers had to squeeze down the face into the cut-hole cavity, the edgers trimmed top and sides, and finally the lifters went off to project the broken muck neatly forward into the drift, where it would be easily accessible to the muckers.

When all was in order, the blaster shouted the traditional "Fire in the hole!" and began to move with all deliberate speed, spitter in hand, from rattail to rattail. The spitter was a length of fuse cut shorter than the shortest rattail on the face and notched down to the core, one notch for each fuse. It had two distinct virtues, the first of which was that the ignition spit from the rattail would not blow it out, as it would a candle, though the miner's candle was used to ignite the spitter itself. The other advantage was its inherent safety feature of singeing the fingers that were holding it about the time one had lingered long enough in the drift. When the fingers grew uncomfortably warm, the blaster automatically threw down the spitter as a preliminary to departure. Later on in the West, carbide lamps would sometimes be used to light rattails—they were reasonably blowout proof—and some mining companies supplied a firework of the sparkler variety which had a limited burning time and which the blaster could carry by hooking its bent end into a buttonhole. But from first to last, miners of the old stock preferred the spitter. Assuming that all went well, the blaster cast a final glance at the face to assure himself that every rattail was smoking and sparking satisfactorily. Then he was free to depart at a measured pace—for it would never do to stumble while running, at the risk of knocking oneself unconscious while in reach of the hurtling fragments of rock.

Having reached a place of safety, he counted the number of reports and made a note of all missed holes (perhaps on a chalkboard) for the benefit of the following shift. If all the cut-hole cartridges fired simultaneously, he would get a count short by two or three of the true count, although normally the initial, extraheavy report thereof would be loud enough to be identified as such. So could the muffled reports of the lifters, muted by the tons of muck now lying across their collars. Occasionally a hole would "miss"; it was usually easy to identify when the men re-entered the working chamber, since a hump would be left on the face where a hole had been drilled but had not carried its share of the work. When air drills came in, missed holes occasionally led to trouble: sometimes the machineman would be drilling from the top of the muck pile (while his mucker removed the pile, benefited by the ventilation from the exhaust of the air drill) and would ram into a missed lifter charge, whose telltale hump was masked by the muck lying over it. Drilling into a miss was a serious thing, but the very muck which hid the missed hole from the driller's view would often have a cushioning effect to lessen somewhat the force of the blast, and he would in addition have the machine between him and the blast to increase somewhat his odds of survival. A working chamber whose walls were heavily dusted with quicklime was the melancholy monument to one who had drilled into a miss without luck.

Because it was the locale of blasting, the working chamber at the head of the drift was kept clear of anything breakable which could not be readily carried away. The stations, or side chambers on each level, usually located near the shaft compartments, were used for storage of supplies and as the miners' lounge. The only continuing ornament of the working chamber was the turning sheet, a large plate of boiler iron which served as a floor covering. Mine cars could be pulled forward onto the sheet from the rails, and its smooth surface made it easier to spot the cars for the greatest convenience of the muckers. A similar arrangement was found on the "grass" at the shaft platform for the benefit of the top lander. In addition, the turning sheet on which the cars were turned about made it far easier to muck with flat-point D-handled scoops off its surface, as opposed to mucking with round-point engineers' shovels off irregular rocky surfaces, as shaft sinkers were obliged to do. When the face was shot out, the broken ore was deposited upon the turning sheet, which suffered little from the process. In some extremely rich gold lodes the turning sheet was supplemented with a canvas, which caught and preserved valuable dusts for recovery—indeed, in such lodes it was economical to sweep down the top and sides of the drift upon the canvas at frequent intervals. The turning sheet and canvas, if any, were dragged forward to assume the same relative position next to the new face as soon as mucking was completed.

The rest followed as a matter of course. The incoming shift sorted and mucked ore and waste, loading each in individual cars to send to the hoist station on their level and to be taken over by the cager for the trip to the surface. Sometimes chutes would carry the ore to a lower level and to the chute tapper, who loaded the cars, or skips, on a demand basis. At the surface the top lander pulled the cars off the cage, replacing them with empties, and then spilled the waste over the dump and took the ore to the sorting shack and bins. From time to time contract freighters brought their wagons to the mine to collect the ore and carry it off millward.

Contract Mining

Occasionally in the American West (though not as often as was the custom in Europe) both development work and ore mining were let on a contract basis; that is, informal syndicates of miners were invited to bid on the total performance of certain specified tasks which the mining company did not care to undertake with its own resources. The job might involve sinking a shaft so many feet or removing so many tons of ore from a stope. Among the Cornishmen the work was organized on a family basis.

The company usually let such a contract only when it had good reason to suspect that the work would be extraordinarily difficult or expensive, hoping that the contract miner would absorb the loss. For his part, the miner was inclined to utilize every economic and technical shortcut known to him, even risking personal hazard or (as often happened) wrecking the workings for those who followed. He would shoot with inadequate powder and short fuses, install insufficient timbering, and perhaps even go so far as to steal and cut sections from the middle of the superintendent's measuring tape, carefully sewing the raw ends together just before final measurement and acceptance. No one really cared for contract mining, as a latter-day song of sarcastic burden implied ("Nothin' could be finer than to be a contract miner in the morning"), but it persisted as an institution partly from economic need and partly from the attractiveness of the battle of wits.

Colorado Lodes

Having organized their prospecting and mining to their satisfaction, the Coloradans fanned out across the Rockies to extend and exploit their discoveries. In the course of the next twenty years they found that there were three distinct and disparate mineral belts in Colorado, each having its own geological and mineralogical peculiarities, and each requiring different mining and milling methods. The Denver District was characterized by gold tellurides carried by quartz veins which cropped out radially from the batholithic domes of the Front Range. It was the discovery of one such lode by John H. Gregory which set off the mining boom of 1859. Tellurides are very difficult to placer—hence the poor placer showings of the Pikes Peak rush. Tellurides were not particularly hard to mine in the firm granite of the domes, but the ore was quite difficult to mill while the state of this art was still in infancy.

The Leadville District of the intermountain park region was found quite by accident when, in 1875, two miners, W. H. Stevens and A. B. Wood, were engaged to cut a ditch to the placer camp of California Gulch, and to their amazement hit silver lead in the stratum of blueish gray limestone which underlay the region. This sort of thing was unheard-of, but the silver lead was indisputably there, and for a time a wild rush of prospectors hammered at every sort of limestone in Colorado. Some investigation revealed that as a rule only the Leadville limestone stratum was worthwhile.[8] The geologists' explanation was that the stratum had been unmineralized when originally deposited, but had subsequently been overrun by a wide white porphyritic flow which did carry mineral. Water action had then leached the mineral out of the cracked and fissured igneous rcck, carrying it down to deposit it in heavy concentration along the contact zone of the "favorable bed" of limestone below.[9] While the tenor of this ore was not particularly high, the silver was found as a chloride (excellent evidence of its redeposition) and the lead as a carbonate quite like the Attic mines at Laurium. Silver values aside, the lead and the limestone matrix were in great demand as flux by the newly opened Black Hawk smelter below Central City, whose operators cheerfully paid premium prices for Leadville ore and concentrates.

[8] Rickard, *A History of American Mining*, 130.

[9] Lakes, *Prospecting*, 171–72.

The third mineral belt lay still farther to the west in the San Juan and Durango area. It had been worked sporadically by the Mexicans for over a century, but now it was subjected to intensive ransacking. The prospectors found placer colors, but for long could not seem to find the lodes from whence they came. At last they learned that they had first to become skilled mountaineers, since the lodes in the Sawatches tended to crop out on the very pinnacles of the heavily eroded, needlelike volcanic necks which abounded in the region as relics of former batholithic boils. This meant that all the necessities for working the mines had to be laboriously hauled (often by tramways resembling modern ski lifts) almost straight up for hundreds of feet, whereupon the miners had to turn about and begin sinking straight down. To make matters even more complicated, this gold ore was heavily pyritic (having been little exposed to atmosphere and ground water in its casing of volcanic rock), and so was as refractory as the Front Range ores to mill.[10]

Milling

Nevertheless, would-be millmen abounded in early Colorado, although they possessed more enthusiasm than knowledge of the art. They drove out of business the hundreds of arrastras which the first-comers had set up but introduced hardly anything except crushing machinery, which could be considered quite an improvement. For instance, the initial shortage of water suggested to many that they could power their mills with John Ericsson's "caloric" hot-air engine, of which the *Monitor*'s designer was for some unfathomable reason very proud.[11] Presently the caloric engines were quietly phased out and normal steam engines introduced; the millmen had discovered that bringing in feed water by flume or even wagon was less expensive than coping with the caloric engine's fuel losses—fuel was expensive, too.

Mills quickly fell into the economic categories of "custom" and "integrated." Custom mills were independent concerns established in districts where no single mine supplied enough ore to support a separate

[10] *Ibid.*, 126, 173–74, 189–90.

[11] Silas W. Burt and Edward L. Berthoud, *The Rocky Mountain Gold Regions, Containing Sketches of Its History, Geography, Botany, Geology, Mineralogy and Gold Mines, and Their Present Operation: With a Complete History of the Mills for Reducing Quartz Now Operating*, 54–55 (hereafter cited as *The Rocky Mountain Gold Regions*).

mill of its own. The properly managed custom mill (at a later date) bought all the ore offered at a price based on assay value, minus a percentage to allow for operation, profit, recovery percentages, and variable ore tenor. It then sold its concentrates to a smelter, or its matte bullion to the nearest refiner. Milling was a speculative business, since miscalculation of the value of an ore shipment could cause either a heavy loss or an unexpected profit. The mill tried to hedge its risk through the discount, but the miner bitterly resented the toll and circulated dark tales of stratagems, treason, and spoils. On the other hand, if a mine or series of related mines had adequate outputs, it was but economic prudence for the syndicate to erect its own milling facilities. Such integrated mills were never too proud to do a little custom work for outsiders when times were otherwise slack, merely in order to keep its machinery and employees at work. Since integrated mills were prone to undercut the custom operators, the practice gave rise to a certain amount of backbiting. As a general thing, the small miner patronized the custom mill, while only the big syndicates could afford to integrate their entire operation from ore to bullion.

The early Colorado mills were run on an incredibly haphazard basis, probably because their operators tried to proceed on gristmill traditions. These mills charged a flat sum per ton of ore delivered to the platform, then contracted to turn back the cleanup to the consignor! This Alice in Wonderland arrangement naturally put no premium on extractive efficiency. There must have been a staggering differential between assay value per ton of ore and actual matte bullion returned, with the millman shrugging his shoulders when the losses were pointed out. Then, too, one wonders how the millmen distinguished between the run of one consignment of ore and the subsequent run from another mine when it came to deciding how much of the cleanup went to whom.[12]

Operations began with delivery of the ore to the mill's ore platform in the narrow, high-sided wagons of three- to four-ton capacity, which were universal in the West. The crusher man worked it down to fist size with a sledge and then tapped it down the chute[13] into a fairly

[12] Lord (*Comstock Mining and Miners,* 114–15) states that this system was used in early Comstock times, while hints in Burt and Berthoud seem to bear out the contention that it was true in Colorado through 1860 (*The Rocky Mountain Gold Regions,* 45).

[13] The crusherman not only broke up the ore lumps but shoveled the ore into

small Blake jaw crusher, which reduced the ore to corn-kernel size, exactly as did its hand-powered miniature version, the assayer's chipmunk crusher. These devices were sometimes called macadamizers, having been patterned upon the machine for metaling roads invented by Eli W. Blake, of New Haven, Connecticut, in 1858. They proved to be great favorites in the West, although some individualists actually attempted (for a time) to use gristmill arrangements whose grinding elements were granite wheels and which must have worn away their corrugations within hours.[14]

Countless inventors attempted without great success to improve upon the Blake crusher, or even to displace it entirely. Despite their attentions, the cast-iron Blake machine persisted nearly to the end of the frontier period, being cheap, light, and effective. Two flywheels stored and delivered the power taken by belt from a line shaft. The crankshaft which ran between the wheels was hooked by a connecting rod to the center of a toggle bolt which greatly multiplied the leverage applied to the movable inner jaw. A small, spring-loaded rod opened the jaw at the end of each stroke. Moving parts were few and rugged; tolerances were easy. Any lump of ore obdurate enough to jam the jaws had to be sufficiently small to be easily cleared by hand from the machine. Virtually all the stresses upon the moving parts were compressive, the one stress in which cast iron is remarkably strong. In short, it was impossible to improve upon the Blake machine without improving it entirely out of the economic reach of frontier millmen (Fig. 62).

From the crusher the ore usually went without screening or further attentions directly to the California gravity stamp mill, which was an improvement of quality, if not kind, upon its Cornish and Spanish prototypes.[15] Universal in the West after 1853, the stamp mill consisted of upwards of five pillarlike stamps whose lower ends, or shoes were replaceable cylindrical masses of iron, each weighing as much as one

the chute. Uncrushed ore will never flow smoothly from hoppers or chutes but seizes every opportunity to hang up and must be punched loose with bars or crows. Thus the occupation of crusherman was hot, dirty, and laborious.

[14] Burt and Berthoud, *The Rocky Mountain Gold Regions*, 54

[15] There was also a steam stamp mill, which was more of a steam hammer, the stamp being connected directly to a steam cylinder. The only one I ever saw was at a mill near the Turning Point Mine, about fifteen miles south of Casa Grande, Arizona (Robert Lenon).

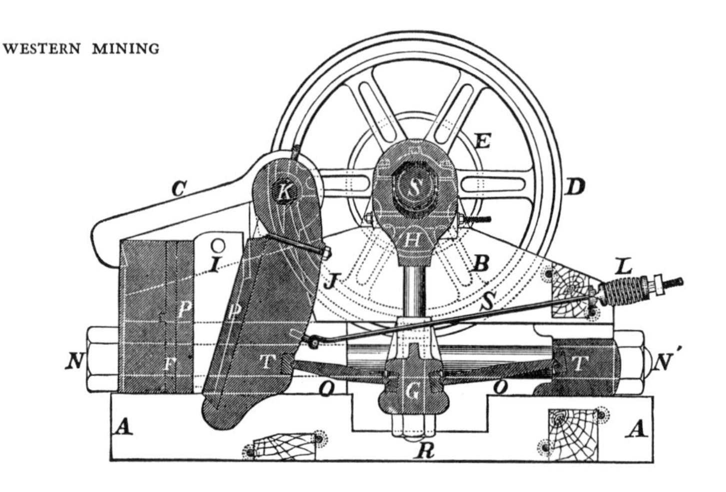

Fig. 62. The Blake jaw crusher. *A–A*, lower timber frame; *B*, upper timber frame; *C*, clamps; *D*, flywheels; *E*, pulley; *F*, main frame; *G*, brushes; *H*, pitman half box; *I*, chucks; *J*, spring jaw; *K*, jaw shaft; *L*, spring on spring rod; *N–N*, main tension rod nuts; *O–O*, toggles; *P–P*, jaw plates; *R–H*, pitman; *R*, pitman rod nuts; *S*, main eccentric shaft. From Thomas Egleston, *The Metallurgy of Silver, Gold, and Mercury in the United States* (1887–90).

thousand pounds. In contrast to the individual square stone boxes of the Spanish *maza*, a heavy iron trough, the battery, ran around and under the shoes, enclosing all of them. In the bottom of the battery, fixed by pouring in molten lead, were equal numbers of flat iron dies, each matching and meeting its appropriate shoe. Provision was made for slotted metal screens in the side of the battery or mortar trough, facing away from the input chute. Overhead, parallel to a sort of mezzanine, ran a substantial drive shaft, equipped with flaring cams, each of which engaged and raised its assigned stamp, then released it to drop a matter of perhaps a foot onto the die. Upon this gangway were sets of hand levers. When engaged, they could lock a stamp in the up position, free of its cam, for inspection or repair. Hanging up the stamps in this manner required dexterity and address, with lost fingers or mashed hands the penalty for inattention (Fig. 63).

Until about 1875 dry ore was admitted to the back side of the battery,[16]

[16] It would have been cleaner and more logical to wet-stamp the ore, but for some reason (probably unimaginative imitation of the *cazo* process), wet stamping was not instituted for a generation, and, it was not until about 1890 that millers

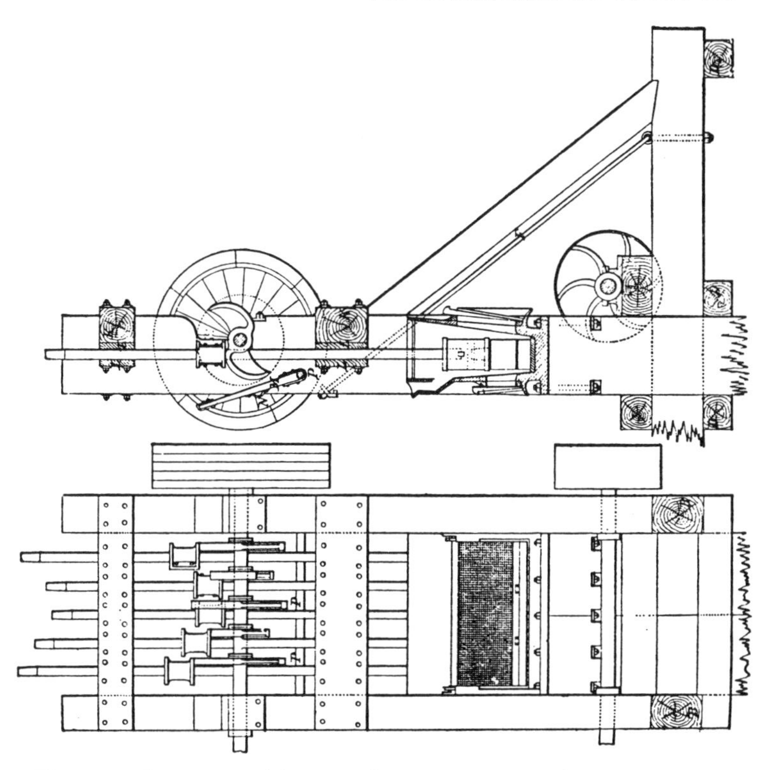

Fig. 63. A five-stamp California mill. From Thomas Egleston, *The Metallurgy of Silver, Gold, and Mercury in the United States* (1887–90).

to make a lateral pass under the stamps and depart out the front side through the screens. The firing order of the stamps was a matter of some concern, since they had to be prevented from nudging the ore sideways and piling it up under one shoe until its stamp was raised out of engage-

discovered the proper way of stamping, which was to introduce mercury, water, and patio reagents to the battery along with the ore. Possibly they were afraid of flouring the mercury or contaminating it with machine grease, but they need not have worried. The shoe is hydraulically buffered on the die so that the mercury does not flour, and cleanliness prevents contamination. In addition, the stamp action is wavelike rather than continuous, mercury preferring the former. In consequence, it is reported that the Homestake Mine operators used the old-fashioned stamps until the 1930's, having found nothing better for their purpose. They designed and cast their own stamps, which, once set up, were peculiarly efficient.

ment with the cam; one order was 1, 4, 3, 5, 2, which kept the ore well distributed.[17] Before the days of automatic demand feeders, the crusher man had to keep a sharp eye upon the battery, for too little ore was even more troublesome than too much; a stamp which came down on a thin layer or none might well knock chips off the shoe or even crack the battery. A properly running mill pulverized its charge, then classified it by spattering the acceptable fines out through the screen. Oversize particles bounced back for another session under the shoes.

Awkward and ear-shattering as they were, stamp mills survived for so long because they displayed much the same virtues as did the Blake crusher: they were readily transportable, comparatively easy to erect upon wooden framing, cheap to operate, stressed chiefly in compression, and repairable even by a journeyman blacksmith. They required no delicacy of handling or construction, and the few moving parts could be cast well enough to obviate any need for fine machining. Mill capacity came to be reckoned by stamps rather than by crushers, a fact which is reminiscent of the old-time cattle inspector who took his census by counting legs and dividing by four. A two-stamp arrangement was estimated as equal to about one arrastra, whereas the largest of the Comstock mills, the California, boasted that it had eighty stamps on line.

Though crushing and stamping were satisfactorily attended to, it is sadly evident that the Colorado millmen were completely at a loss where to go from there with their refractory ores. Most seem to have decided upon simple amalgamation, with either the pan or the *cazo* process, whose only effect would be to lose up to 75 per cent of the assay values. One of the best mills, the Black Hawk, ran the stamped and classified dust over bumper tables and a vanner to concentrate the ore, then introduced it, together with salt and mercury, to Freiburg amalgamating barrels, which were the crude prototype of modern ball mills.[18]

[17] This firing order was deduced from the diagram of a stamp-mill line drawing in Lord, *Comstock Mining and Miners*, 234, although in the diagram cam 2 seems to have an ear omitted.

[18] Burt and Berthoud, *The Rocky Mountain Gold Regions*, 54. The authors seem a bit hazy on the way a vanner operates. The ball mill, which became a standard feature on the western mining scene after 1900, consisted of a large steel cylinder or drum which was slowly rotated about its long axis by means of a gear train. Inside were large numbers of forged iron balls, up to three inches in diameter (today up to five inches). Crushed ore and water (and other matter) were put

The Freiburg process was an effective combination of chlorination and amalgamation which economized on both fuel and mercury, though at the expense of requiring more elaborate machinery than either the patio process or *cazo* treatment of silver sulphurets. The ore was cobbed to eliminate waste, crushed in a Chilean or Blake machine, stamped to fine flour, then mixed with 10 per cent of its weight in common salt and roasted in a reverberatory furnace. The roasting drove off sulphur, arsenic, and other undesirable elements while converting the sulphurets to chlorides. The roasted ore was then introduced in five-hundred-pound batches to barrels of one-ton capacity, designed to rotate around their long axes at the impulse of water or steam power. Water was run in, and the barrels were rotated slowly until the ore was a paste. At that point about fifty pounds of mercury and quantities of copper or iron salts were added to reproduce the final reactions of the patio process. The barrels were then run for upwards of twenty-four hours. Then they were stopped, and the agglomerated amalgam was tapped down from the barrel and removed for retorting. The remaining contents of the barrel were run into vats and agitated in water to work off the tailing, leaving behind additional amalgam for recovery. Millmen were delighted to find that when the Freiburg process was properly run it cost them but seven ounces of mercury per pound of silver recovered,

in the cylinder. As it turned, the balls rode up the inside and then fell back to pulverize the ore charge and mix it with the other ingredients. (In modern mills the cylinders are lined throughout with corrugated steel shoes which are easily replaced when worn thin. Millmen estimate that they lose one pound of ball and one of shoe for every ton of ore ground.) Ball mills are used with dry charges for cement making, but with wet charges for ore grinding. They instantly won the day everywhere but in the precious-metals field, being cleaner and more efficient in almost every respect than were the archaic stamps. The gold and silver millmen, however, regarded them askance; they were very prone to flour the mercury in their charges, they were continuous and not wavelike in their treatment, and they pulverized a percentage of the ore too finely for the extractive processes then obtaining to work at their best.

To overcome these objections, rod mills were introduced. They were like ball mills in every respect save that their grinding elements were full-length steel rods in diameters up to three inches. The rods theoretically could come no closer together than the size of the largest fragment of ore separating any two; therefore, they selectively crushed the large pieces, then the next-largest, and so on until most attained a uniform and optimum size. When less than infinitely small particle size is desired, rod mills are ideal; but ball mills still predominate in some industries, for example, the copper industry.

in contrast to the patio process, which lost a pound of mercury for each pound of silver gained.[19]

Sometimes the millmen ran their classified dusts over bumping tables, which were broad, flat surfaces equipped with low riffles and amalgamation catchplates, the whole being hung at a slight angle to the horizontal from flexible straps or pivoted bars. Plain or amalgamated mill pulps went in the high end and came out the lower end after being jigged about over the riffles and plates. Motion was given the table by means of a connecting link, which ran to a power-driven eccentric, although hand power would have sufficed to swing one in the one-way fore-and-aft jolt required. The bumping table may be regarded as a poor hybrid of sluice box and cradle, a prototype of the shaker or concentrating table to come but having the aggravating habits of all three and no particular virtues of its own. In consequence, it died out very rapidly.

The concentrating table, which quickly replaced the bumper, was more acutely inclined, vibrated continually in a back-and-forth lateral motion, and was equipped with shallow *vertical* riffle bars which terminated before lower-end exit ports. The pulps and a trickle of water were led over one corner of the upper end, from which they flowed down the surface. As the table "snapped," the gangue was urged sideways across the riffle bars, whereas the concentrate or amalgam remained on its own side, ultimately dropping through the exit port. A little reflection makes it evident that concentrating tables were more productive than bumpers, having the additional advantage of cleaning themselves automatically and thus eliminating periodic down time for amalgam removal.

Crude sampling techniques were used to evaluate the process as it progressed. With a special curved spatula a specimen of amalgam was removed and carefully washed clean, and its condition was scrutinized. If the mercury globules were bright, ran together readily, and were soft, either the ore was lean or else the batch required more chemical digestives, which at that time meant common salt and little else. If, on the other hand, the globules were crisp and had a frosty appearance, they were fully charged with gold and silver. In that case, it was time to

[19] H. E., "The Reduction of Silver From its Ores," *The Weekly Arizonian* (Tubac, Arizona), Vol. I, No. 22 (July 21, 1859).

clean up the machine and put additional mercury into the amalgamating barrel.

From bumper or concentrating table (which could be used either before or after amalgamation) the depleted pulps were next led to a cascade of amalgamation pans, changed very little from those invented by Velasco. These were arranged stairstep fashion so that the pulp of the highest pan would overflow down into the next, and so on. The only latter-day innovation was the inclusion of mechanical paddles which slowly stirred and agitated the pulps with the intent of bringing each particle sooner or later into contact with the mercury film which coated the pan's copper interior. Even so, before chemical leaching was devised, much of the amalgam passed out and away. When times were slack, the mill ran its screened tailing over the tables and through the pans a second time on the theory that whatever cleanup of gold, or even mercury, was made beyond the cost of fuel and labor represented virtually unearned increment.

To move ahead a bit, it might be said that in the days before flotation and cyanidation came of age, the Wilfley table was regarded as the capstone of all concentrating technology. Its origins clearly lay in the old concentrating table—and doubtless in the observation that heterogenous mineral concentrates did not pile up randomly behind the latter's riffle bars but rather that a higher percentage of the heavier minerals were trapped behind the nearer bars in uniform proportion to their occurrence in the ore. The lighter minerals, pyrites, for example, tended to be found behind the farther bars. All one had to do, therefore, was to adjust matters so that the separation was clean-cut rather than statistical, each mineral being trapped behind its assigned riffle bar and then marshaled away to its planned destination. The trick was not easy—even to describe—but a Wilfley table dancing its hula-hula was by all accounts fascinating to watch, as each streak of mineral piled up at its predestined bar, only to be jigged away in a steady band to its particular exit port. Yet the Wilfley died out, being both temperamental in itself and more expensive than differential flotation. Nevertheless, in its day, the Wilfley was the answer to a concentrator's prayers since it increased substantially the side profits from metals which had hitherto been merely dumped as tailing or lost in slag.

Still and all, comparison of ore assays with bullion recovered made it painfully apparent to the Colorado mining syndicates that the greater part of their values was getting away from them, or, if you please, that neither straight amalgamation nor the *cazo* process was doing more than skimming off the cream of their ore. A stopgap solution was to concentrate the ore in Colorado, then ship the bagged concentrates overseas to Swansea, Wales, for smelting and refining. Swansea had the experience and the best-quality Welsh steam coal, but the costs of such shipment before construction of the Union Pacific Railroad were painful. Virtually every mine prospectus of the early period mendaciously claimed that recovery of the base metals present would pay in full for the long trip to Wales, leaving the gold and silver values as pure profit. Only the inexperienced were deceived, and then not for long; Colorado mining slumped badly for a time.

The first step toward salvation came with the discovery of fields of coking coal in the sedimentary strata of Colorado. In 1868 the Boston and Colorado Smelting Company opened its shop at Black Hawk to take advantage of the possibilities. It did moderately well for a decade, but the discovery of the silver-chloride and lead-carbonate ores of Leadville really put Colorado smelting on its feet, since the Leadville limestone provided in one package all the ingredients still missing from the process. At that point the innumerable amalgamation and *cazo* mills either converted completely to concentration alone or else went quietly out of business altogether.

The smelting of refractories much resembles a three-ringed circus, in which quite a few acts are in progress all at the same time. High heat to provide the energy necessary for chemical dissociation of the metal salts is essential; it is provided by the coke and the air blast. But these requirements in turn demand a furnace which is either water-cooled or lined with firebrick or both, neither cheap to build nor easy to keep in repair. The air blast supplies oxygen to combine with the sulphur present and remove it as a stack gas of sulphur dioxide; it also carries off tellurium and arsenic in the same manner. However, the surplus oxygen constitutes a problem in itself and has to be dealt with somehow. Fortunately, the combustion of coke provides carbon monoxide in great quantities, and carbon monoxide likes nothing better than free oxygen with which it can combine to form carbon dioxide, which likewise departs by way of the stack.

The lead carbonates which have been included in the charge are dissociated quite readily into lead, carbon dioxide, carbon monoxide, and oxygen (all of which are either useful or innocuous). The lead combines with the silver present to form a low-melting-point alloy or descends in droplets through the charge to carry down the gold, exactly as in fire assaying. If there is copper present, it comes down, too, and must be refined away later. This molten matte is tapped off at intervals, run into molds, and held for further attentions at the hands of the refiner.

We are now left only with the problem of the slag. Slag is iron silicate (iron from the pyrites present, silica from the quartz) whose melting point is so very much higher than that of the nonferrous matte that it floats up as a clinker, accumulating with it all the chemical odds and ends—calcium, magnesium, sodium, and whatnot—which entered the charge at the beginning of the run and are now rated as undesirables.[20] The slag is skimmed off from time to time and dumped. All that is now needed is the supervision of an industrial chemist to diagnose the requirements of the particular concentrate charge to be smelted and to prescribe appropriate amounts of pyrites, lead ore, quartz sand, and limestone (if dolomite is available, it is substituted for limestone). When well done, smelting ends with gas, liquid, and solid (stack gas, matte bullion, and slag), with nothing left over.

[20] Railroad construction engineers in the West quickly found precious-metals slag an ideal ballast material. They used it by the carload and praised it highly, to the consternation of their eastern colleagues, who had found iron-furnace slag good for nothing. The iron-silicate precious-metals slag was an extremely tough and durable ceramic, whereas open-hearth slag was composed of light-metal silicates whose melting point was lower than that of the iron from which it had been removed and whose physical characteristics left much to be desired. Even natural iron-silicate dikes are tough enough to wreck steel tools, as I discovered at the cost of a brand-new geologist's hammer.

CHAPTER VIII

Developments Sacred and Profane

WITH THE BEST OF GOOD WILL, American frontier mining before 1870 can hardly be called much of an improvement upon the Saxon methods described by Bauer three centuries earlier. This regressive trend was due partly to the crudities of the Spanish tradition, partly to the lack of trained engineers and of colleges of mining engineering, and partly to the universal frontier attitude that waste hardly mattered since resources were unlimited. By 1870, however, there was cause for each of these factors to be modified, and nonferrous mining and milling began to experience something of a technological revolution in quantity, if not in kind. The Industrial Age was making its impact felt, chiefly through the use of iron as a material, through steam for power, and through reciprocating-machine design. This revolution was perhaps most marked in the Comstock Lode, whose technical origins were on a par with the most primitive methods of the Spanish tradition (indeed, the Mexican Mine of the lode was Mexican-owned and was still employing *barreteros* and *tenateros*, while the other mines were moving toward machine drilling and elaborate steam hoists) but which closed out as a showplace of the Iron Age. For the sake of clarity, however, these changes will be treated individually and will not be identified with any particular mine or mining district except in the case of less typical or more striking illustrations.

Drilling and Blasting

Perhaps the first improvement was the revolution in drilling and blasting. As early as the American Civil War, European and northeastern United States railroad-tunnel engineers had begun to experiment with steam drills. These devices held great theoretical promise of speed and economy, especially in supplanting brutal and expensive double-jacking, and they were given many trials. But John Henry was right—they were at best only an expedient. The boiler had to be erected outside the adit or shaft and the steam conducted most of the way to the face through insulated iron pipe. Despite the insulation the steam lost a great deal of energy during the trip; and every time the heading was advanced, the system had to be cooled down or expensively valved in order to be extended. No one was able to come up with a really satisfactory flexible steam hose to connect the power line with the drill; all flexible hose leaked and blew out to a remarkable degree. The miners and tunnelers found the wet heat of the exhausts intolerable. All agreed that the idea was a good one, but impracticable.

Though theoretical efficiency was much reduced,[1] mechanical drilling became a reality with conversion of the steam drill to compressed air. There was suddenly a concerted rush to pneumatic drills. Compressed air could be conducted almost infinitely far without need for insulation and without excessive loss of energy; flexible hose for compressed air presented no insoluble design problem; and the exhaust at the face was cool and invigorating, if somewhat tainted with lubricant.[2] What a mining superintendent lost in fuel inefficiency, as compared with steam drills, he more than made back in lowered ventilation expenses. The initial investment was modest: (1) a compressor which ran off a line shaft from the main steam plant, (2) low-cost common piping, and (3) cheap, rugged Burleigh, Rand, Leyner, or Ingersoll drills, far simpler in design than their European counterparts, which had a profusion of gadgets and which were in the shop for repair more often than they were at work in the heading.

The drill itself had the basic design of a double-acting steam engine, with the piston connected to a drill chuck instead of a connecting rod. Air was valved into the rear half of the cylinder to slam piston and

[1] About 90 per cent of the energy used in air compression is wasted as heat loss.

[2] Lubrication of compressors and drills proved quite a problem, and many a miner was killed by the partial combustion of oil of unsuitable quality; now and then the high compression of oil fumes would make an explosive mixture and blow up the compressor itself. I have always thought that the Diesel engine thus invented itself, as it were (Robert Lenon). Lest scientific historians take umbrage at this comment, it should be noted that Rudolf Diesel experienced no flash of insight while observing an air compressor sail skyward but rather approached the principles of his engine from an entirely theoretical basis and only during the course of his experiments tried out liquid fuels. Cf. T. A. Burgess, "Explosion in Compressed-Air Main," *Mining and Scientific Press*, Vol. XCVII (August 22, 1908), 253.

attached steel forward. At the top of the stroke the piston extension at the rear tripped an automatic valve which bled a brief shot of air into the front half of the cylinder to return the ensemble to the bottom of the stroke. At that point the valve was tripped the other way, and again piston and steel shot forward. Piston and drill steel were rotated automatically during strokes by a rifle bar, while the whole assembly was advanced into the hole by a hand-crank-operated jackscrew at the base (Fig. 64). Drills proper were mounted by an adjustable universal joint upon an iron pillar, called a column or bar, which ran between floor and roof of the drift or horizontally—when lifter holes were being drilled, they were supported by a plank trough, the Finn board (Fig. 65). Hand-held versions, very like a modern jackhammer in appearance, were used in places where pillar support was not feasible. These drills sounded like demented woodpeckers, but made hole in an incredibly short time.[3] A mine could lay off three-quarters of its double-jack teams, yet still hoist more tonnage than ever it had before. Supervisors regarded

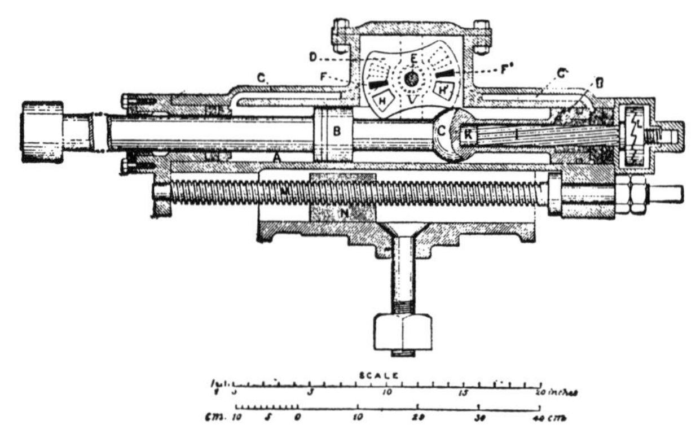

Fig. 64. A Climax machine drill. *B*, piston; *C*, piston extension, actuating air valve; *I*, rifle bar; *M*, screw jack; *V*, air-valve cam. Chuck is at the extreme left.

[3] They also dulled the primitive steels in an incredibly short time. A mine had to hire nearly as many blacksmiths to keep up the supply of steels to the face crews as it laid off double-jack teams. Although such improvements as high-alloy steels and detachable drill bits came along in due course, the insatiable gluttony of air drills was not really met until the post-1945 development of tungsten carbide-tipped tools.

Fig. 65. The Finn board. Reproduced by permission of Buck O'Donnell and Shaft and Development Machines, Inc.

the machine drill as the best thing since the invention of gold itself. The miners, however, were not so certain. They suddenly seemed to be dying in droves.

Georg Bauer had warned them,[4] but few people in the West at first fully appreciated this sequence of events. The miners knew only that men died unpleasantly when air drills were being used in hard rock, and that the killer made itself known by a hard, racking cough which so much resembled the then-all-too-prevalent pertussis of tuberculosis that their ailment was called miner's consumption. Physicians of the period were baffled, though even a tyro could tell that the deceased miners' lungs were in very poor shape, appearing to be almost petrified, but that they did not have the peculiar ping-pong-ball pathology which

[4] "This dryness causes the workers even greater harm, for the dust which is stirred and beaten up by digging penetrates into the windpipe and lungs, and produces difficulty in breathing. . . . If the dust has corrosive qualities, it eats away the lungs and implants consumption in the body; hence in the mines of the Carpathian Mountains women are found who have married seven husbands, all of whom this terrible consumption has carried off to a premature death" (Bauer, *De Re metallica*, 214).

characterized pulmonary tuberculosis. The etiology varied significantly as well. Miner's consumption lacked the fluctuating temperature, hectic flush, and night sweats which were so typical of tuberculosis. Ultimately research made it manifest that the culprit was silica dust,[5] churned up in huge quantity from quartz, porphyry, or granite by the new drills. Under the microscope the dust was shown to possess murderously sharp edges—the miners had been breathing in millions of Lilliputian razor blades. This dust scarified and altered the lung tissues until the miner literally smothered to death. Some checking showed that stoneworkers everywhere who dealt with silicious dust or rock with power instruments were prone to die of silicosis. The remedy seemed obvious: prohibit the use of air drills and go back to double-jacking.

Those, however, were the days of Associate Justice Stephen J. Field, who was leading the United States Supreme Court in the formulation of economic doctrine so reactionary as to make Cato the Censor appear a liberal by comparison. The Court met the state regulatory laws against the "widowmaker" drills by decreeing that if a man accepted a job in a hazardous industry, knowing or suspecting that it was hazardous, his blood be on his own shirt front when the hazard caught up with him.[6] After all, the Court reasoned, the employee had not been forced to take the job; by the same token, the employer's legal and moral responsibility began and ended with the weekly pay envelope. Nor was the corporation under the slightest obligation to abate the hazard if that abatement cost it money. State safety regulation, continued the Court, could not tend to deprive corporations of their property without benefit of a due process, which the Court was prepared to interpret very narrowly. Indeed, then as now, due process of law was nothing the Court

[5] Silica dust is not dissolved by body fluids in the manner of the dusts of limestone or other basic rocks, and hence tends to accumulate in the lungs as well as to inflict traumatic injury. Very recent research hints that there is even more to the pathology of the disease, for the insoluble dusts of other chemicals (such as diamond) do not produce the petrification effect, which is probably a product of the silicic acid released in the tissues. In addition to the problem of drilling, exposure to silica dust was pronounced in mucking, chute tapping, stoping and crushing—everywhere, in fact, that dry ore or waste was handled in large quantities (Robert Lenon).

[6] The actual cases in litigation included *Allgeyer v. Louisiana* and state railroad safety-equipment laws, but the Court's reasoning could be and was directly transferred to state mining-safety regulation.

denied it to be, excluding anything the Court just disliked the idea of. In conclusion, therefore, if a miner wanted to quit or to provide his own protection against silicosis,[7] he was at perfect liberty so to do; it was still a free country. Dry drilling with its attendant white cloud of death went merrily on.

Western-state Populist agitators and organizers of the Western Federation of Miners fought the situation, but it was the technicians who came up with a universally satisfactory solution. About 1890, J. G. Leyner, of Denver, hit upon the idea of mucking out drill holes by the simple device of diverting some of the compressed air down a longitudinal hole in the drill steel to blow away the rock dust as fast as the bit created it. This was fine for the steels, but miners quite reasonably refused to work with his air-blown machine. Leyner took further thought and redesigned his drill so that water instead of air went down through the hole. The water would not only create a mud to act as an effective wet-grinding compound but would flush out the hole as well, even in down holes. Incidentally, it also cooled the steel and wetted down the flying clouds of silica dust.[8] In the granite mills of the East they tried trickling water over the cutting saws and found that it worked equally well. All the mining and quarrying states promptly enacted anti–dry-drilling laws, and since dry drilling no longer paid, the companies forbore to appeal.

Leyner's water-flushed drill imposed many problems which had to be solved before it could be sent down into the mine. Drifting a one-eighth-inch hole for six feet with reasonable truth through a solid steel bar is easier said than done; it is reported that for his first experiments Leyner himself used a length of common iron pipe with chisel bit welded to its tip. This fragile expedient would not stand up to the stresses imposed upon it, and in coming up with something better, the steel industry had to break new ground in its forging and machining processes. Likewise, it was evidently unfeasible for the flushing hole to terminate in the middle of the chisel bit. This difficulty was temporarily overcome

[7] This protection was gained in part by casting water into the flat or down drill holes from a tomato can. Drilling was carried out by a two-man crew, a machine-man and a chuck tender, and while drilling was going on, the chuck tender was forward of the drill, against the face, and had no duties beyond the arrangement of the next steel or two. Thus he was free to pitch in the water now and then.

[8] Sandström, *Tunnels*, 294.

by putting the exit port at the side of the steel on the end of a short down slant from the central axis hole (I have seen specimens of this steel at the Vulture Mine). Such an arrangement weakened the steel, however, and it was not long before bits were redesigned about the star pattern in which the port is in the center, with four short individual bits radiating from it (Fig. 66). A generation later a detachable star bit was invented by a foreman at Butte, Montana. Drill steels are roughly dated with some ease: plain chisel-bit steels predate 1897, the year of Leyner's patent; the machine chisel-bit steels with forged chuck ring and side flushing port lasted about ten years; the plain star-bit steel with center flushing hole persisted to 1930; steels with detachable bits are relatively modern. All sorts were used without modification in either

Fig. 66. Early-day stoping. Reproduced by permission of Buck O'Donnell and Shaft and Development Machines, Inc.

Fig. 67. The wiggle-tail (self-rotating) stoper. Reproduced by permission of Buck O'Donnell and Shaft and Development Machines, Inc.

horizontal drills or vertical stopers,[9] although the latter frequently dispensed with water flushing in up drilling (Fig. 67).

[9] A stoper was a columnar drill used for up holes in stopes. Its top end was a conventional drill; the lower end had what would now be called an air leg, a foot driven by the air in opposition to the drill, which not only afforded a firm foundation but was extruded as the drill advanced to keep it well up in the hole. It lacked a rifle bar and was rotated by a side bar turned by the miner. In action it resembled a mechanized pogo stick.

The second technical revolution of the 1870's came with the introduction of high explosives. Before the Civil War blasting was done with gunpowder compounded of the archaic saltpeter-sulphur-charcoal formula. Between 1866 and 1875 the miners used Du Pont "soda powder," which substituted the less-expensive sodium nitrate for the old potassium nitrate but otherwise was little or no better at pulling rock. Then came nitroglycerine, which seems to have been initially used in the West by the Central Pacific Railroad for clawing up through the Sierra passes. Since nitro was easily five times as powerful as black powder and had a good deal more shattering effect on hard rock, it attracted attention despite its evil tendency to explode if one merely looked cross-eyed at it. In addition, it froze, decomposed spontaneously, could not be readily inserted in up drill holes, and was almost impossible to transport safely any distance. Only big companies could use it, since this devil's broth had to be manufactured near the site of use at a considerable cost in equipment and (all too frequently) overhead. The average miner eschewed nitro; any fool could home-brew it, but only another fool would attempt to use "blasting oil" that had not been carefully kept ice-cold during the reaction period, washed, and deacidized. There were a few such fools to be found at first, but their numbers rapidly declined.

Alfred Nobel tamed nitro by adding it to inert fillers such as sawdust. His new dynamite had nearly the shattering power of pure nitro but was infinitely easier and safer to handle. It was still prone to freeze, but the cautious man could thaw out the new "giant powder" without blowing himself up too frequently.[10] Precisionists could alter the nitroglycerin percentage to match the toughness of the rock in question. One could load it in up holes, tamp it down lustily, and shape charges to meet any contingency; but best of all, it would stand for a great deal of the mistreatment normal to rough-and-ready mining. Presently the wooden dynamite case entirely replaced the old powder keg as the miner's chair and table, and westerners were using dynamite for a thousand purposes which Nobel had never dreamed of. It would dig irrigation ditches, fell trees, discourage claim jumpers, or assassinate

[10] In the old days dynamite was thawed by burying it in several inches of earth, building a small fire nearby, and then stirring embers over the burial site with a stick until the dynamite was warmed and ready to go. It was considered wise to devote one's entire attention to the task and not become distracted by other activities. Today glycol antifreeze is sensibly included in dynamite.

former governors of Idaho. It was said—not altogether in jest—that a good blaster could blow a man's nose for him without mussing his hair. Finally, it became a common means of suicide among prospectors suffering from real or alcohol-induced discouragement.[11]

Like the parent nitro, the fumes of exploded dynamite were head-splitting—even suffocating in heavy concentrations—but the chief hazard of the new blasting lay in the small copper detonators which were essential to setting it off. The fat paper candles were so inert that a preliminary shock of some sort was necessary to detonate them. This shock was provided by a pinch of fulminate of mercury or some other supersensitive explosive packed in the bottom of the innocent-looking little copper tubes. In practice, one thrust a Bickford fuse or electric-resistance wire well home in the detonator, crimped its neck with a special plierlike tool (fools were prone to use their teeth for this purpose), and inserted the ensemble in the hole punched in the side of the dynamite stick. The whole was bound firmly together with a sling of coarse, jutelike string, put into the hole, stemmed well, and then shot. Detonators, or blasting caps, were small and useful but vindictive when abused. It was considered well to transport them in special padded and softly sprung mine cars[12] and not to carry them loose in the pockets or to store them adjacent to dynamite.

Before these precautions were adopted, caps were scattered about promiscuously, with predictable results. William Wright held forth on the consequences by noting:

> Persons unacquainted with their uses always appeared to be overcome by an ungovernable curiosity in regard to the nature of their contents the

[11] Roscoe Willson, "Arizona Days," *The Arizona Republic*, December 31, 1967, 23.

[12] The Mining Museum at Jerome, Arizona, has on display one of the detonator-transport cars used in the United Verde Mine. Adjacent to it is another mine car of equally novel pattern, though unlabeled. Its body is a low metal box set parallel to the rails. The top of the box has three annular apertures, each about twelve or fourteen inches in diameter and spaced equidistantly apart. The apertures are individually covered with hinged, circular wooden lids, so contrived as to swing back when not covering the openings. The observer may be deceived into fancying a functional resemblance to those rural systems described in *The Specialist.* I took another tack, however, and asked the proprietor of the museum whether the car in question might have been used for the collection and transportation of high-grade specimens, and was instantly and emphatically informed of the correctness of my surmise.

moment they, by any means, get hold of one of these caps. The first thing they do is begin probing and scratching in the interior of the little cylinders in order to get out and examine a sample of their contents. It invariably happens that at about the first or second scratch the cap explodes, and the person engaged in prospecting it loses the ends of two fingers and the thumb of the left hand.

In Virginia City and Gold Hill about one boy per week, on an average, tries this experiment, and always with the same result. . . . Miners very frequently carry these caps loose in their pockets, often mixed with their tobacco, and thus, occasionally, get them into their pipes. Several favorite meerschaums have been lost in this way, and the ends of a few noses.[13]

The coming of air drilling and high explosives made possible the opening of medium-sized, relatively low-grade gold and silver mines of the intermediate period, particularly those of California, Arizona, the Great Basin, and the northern Rockies. Up to about 1875 small lodes were worked by one or two men who did everything from hoisting ore by windlass to their own arrastra milling. Anything bigger had to be so big and so rich as to cover a tremendous payroll of dozens to hundreds of double-jack teams working two shifts. After 1875 a mine could be opened and worked profitably with a payroll of two or three dozen men, hoisting enough ore of modest grade to net, what with the government subsidy on silver, a comfortable dividend on the operation. The economic and social consequences were instantly apparent: hundreds, if not thousands, of medium and medium-small western mines were opened in the period 1875–93. In Arizona alone thirteen thousand "mines" (more accurately, claims on which minimal location work had been done) were recorded by 1876. Within a decade the number doubled, but substantial development was confined to about seven hundred in the one territory alone. Thus all over the West there grew up a large number of small mining communities, the population of each working in the mine and the concentrator or mill or serving the community as a whole.

Transportation

In mine transportation such truly revolutionary developments as block-caving, "electric mules," and the like were undreamed-of. There was

[13] Wright, *The Big Bonanza*, 148.

only steady improvement of existing methods. The whim and windlass gave way to small, inexpensive steam hoists which, like the proverbial Yankee whaling brigs, were apparently produced by the mile and cut off by the yard. Hemp gave way to Roebling's steel wire rope. That accounts for the surprisingly large circumference of the headframe sheaves, for, whereas hemp will readily run over a sheave of small radius, wire rope must be led gradually around an arc of fairly great radius. To do otherwise would ultimately entail deplorable consequences to anyone standing behind the hoisting drum or upon the hoist cage at the moment of failure. The highest development of hoisting rope resulted in the use in the Comstock mines of custom-made braided steel wire rope (actually, flat belts) which not only could handle ten tons with ease and no danger of slippage on the hoist drum but were gradually tapered toward the cage end, so as to reduce the dead weight to be hoisted. The flat rope and reel gave an excellent mechanical advantage on starting as well (Fig. 68).[14]

In developed mines the good old sinking bucket was always replaced by the cage, a simple platform which was held in alignment by a pair of wooden rail-like guides running down the shaft. Miners were for long uncertain about the virtues of the cage—they complained that it was incapable of raising water from the sump when otherwise unemployed, but there is a suspicion that they really disliked the lack of any sort of sides or overhead enclosure on the cage. On the other hand, men who did not have to ride in a cage pointed out that it did not spin about as a bucket did, that the bucket was exceedingly awkward to clear of ore and waste, and that the bucket's accommodations were very limited. At length someone had the inspiration to enclose the cage with wire mesh and to add a safety brake to ensure against unscheduled drops. The miners accepted this "safety cage" wholeheartedly.

As with hoisting, there was virtually no change in mine pumping except steady enlargement of the Cornish pump (see Plate 8). The growing costs of this clockwork and its fundamental unreliability produced a rash of experiments in other directions, but all fell short on the basic design problem of getting power down to the sump, where it was to be applied. That could be done only by mechanical transmission (pump rods) or by steam lines (whose shortcomings have been discussed). One oddly

[14] Lord, *Comstock Mining and Miners*, 224.

Fig. 68. A shift going underground. Reproduced by permission of Buck O'Donnell and Shaft and Development Machines, Inc.

successful variation was an ingenious water lift, put in by the great Consolidated Virginia Mine toward the end of its working life. The Con Virginia had reached the 2,500-foot level below its engine house, far below the effective reach of pump rods and several hundred feet below the heading of the Sutro Tunnel, which had been draining the mine by gravity to that point. The engineers took thought and observed that the Virginia City water supply was brought from Marlett Lake, high in the Sierras to the west, taken by inverted siphon up to the peak of Mount Davidson overlooking the city, and from there distributed. They tapped the line at the peak, building a penstock which led down nearly 4,000 feet into the sump of their mine, and got a tremendous head of water at nearly 1,500 pounds per square inch. This formidable pressure was then diverted upward into a bell nozzle sunk beneath the water level in the sump, and it sucked the mine water up far enough to be discharged into the Sutro Tunnel drainway. This penstock pressure was so great that the jets from even small leaks drilled through timbers like a bullet. The miners picked up sums of money betting visitors that they could not cut through these jets with an ax—when the experiment was tried, the ax would be violently twisted out of the unwary better's hands.[15]

Crushing and Stamping

Inventors of the period were forever attempting to improve on crushing and stamping methods—so much so that it would be impossible even to attempt a list of the various patents, let alone to describe each new machine. Most were merely variations on a few standard devices, such as Scoville's Patent Premium Quartz Mill, coyly advertised as "unquestionably superior in every respect to any Stamp Mill," which turned out upon inspection to be merely a geared, cast-iron version of the old Chilean mill.[16] The Huntington Centrifugal Roller Quartz Mill was out of the same stable, the axes of the crushing rollers being set vertically instead of horizontally. For their part, quartz millmen continued to prefer Blake jaw crushers and California gravity stamps until something indisputably better came along.

The something better did not put in an appearance until the mining frontier was nearly played out. It was the cone crusher (sophisticated versions of which are the backbone of today's very-low-grade-ore copper industry), patented by P. W. Gates in 1881. Considerably more efficient

[15] Henry Curtis Morris, *Desert Gold and Total Prospecting*, 35–36 (hereafter cited as *Desert Gold*).

[16] Burt and Berthoud, *Rocky Mountain Gold Regions*, unpaged advertisement.

than the jaw crusher of whatever size, it was continuous and rotary, rather than discontinuous and reciprocating in its action. A massive iron cone hung point downward from a sort of arch. At its lower extremity an eccentric gear joined it to its power train. When in operation, the cone was impelled to wobble about, slightly off center. Surrounding the cone was a somewhat more flaring fixed cone; at one point the opposed walls were nearly touching, and at the opposite point they were the maximum distance apart. Ore was dumped into the bell-like top flare and the machine set going. As the cone wobbled about, there was much tumult but seemingly little activity (the wobble was too slight to be obvious), until the observer noticed that the blocks of ore were mysteriously cracking, falling in fragments that dropped farther down, cracked again, and finally vanished into the dusty dark below. The repeated waves of constriction between wall and cone had rather neatly done in the ore.[17]

The source of all such activity reposed down at the other end of the mill, in the shape of the classical double-action steam engine, which changed not the slightest bit (except in size) from the beginning to the end of the American mining-frontier days. It was crude, rugged, cheap, and reasonably portable in sections, and it had an efficiency curve so ludicrously low as to be almost unbelievable. Nevertheless, the firebox could digest any fuel from coal to greasewood brush, and anything was made do for boiler feedwater that could be pumped in.[18] The boiler might grandly be described as of the fire-tube variety; in actuality it was but a simple boiler-plate shell, pierced by a few flues which served to conduct the hot gases of combustion back through the boiler once they had warmed its nether surface. Boiler accessories might but did not

[17] Sometimes a particularly obdurate lump jams the cone. In that case the crusher man employs a light air drill to drill and shoot the offending lump *in situ*, the heavily built crusher suffering not at all from this measure. If time presses, he can make an "adobe shot" such as those made by roadmakers or prospectors who wish to investigate the interior of an intriguing boulder. He merely lays the prepared charge atop the rock, poulticing it well with damp clay. The process fills the air with adobe but usually cracks the rock to the satisfaction of the operator.

[18] The railroads, which were fastidious about the feedwater used in their locomotives, occasionally bypassed eligible division and watering points solely for the reason that only unsatisfactory (corrosive) water was available. Mills and concentrators, however, made do with what nature provided, since engine breakdowns (meaning, usually, boiler failures) are mentioned with great frequency.

necessarily include such esoteric refinements as condensers, pressure gauges, and safety valves. A smokestack of sheet iron and of unpretentious height completed the apparatus. Any blacksmith could botch together such repairs as were required. The violation of elementary safety practice was so common as to constitute standard procedure. The rule of thumb was that the water tender either ran his boiler well below rated pressure and played it safe or else crowded it for all it was worth and took his chances. Either way, a man knew what he was doing and could conduct himself accordingly; but with fuel as expensive as it was in the West, it is likely that few mining superintendents or millowners encouraged high pressure.

At working pressures of perhaps fifty pounds the steam was collected in a dome and from there sent through a "dry pipe" and past the throttle to the engine, whose valves, cams, and rods would have been perfectly familiar to James Watt. Three systems were in general use. Small-mine hoisting machinery might be of the vertical-boiler donkey-engine variety, whose minute cylinder either was directly coupled to the hoist gearing or drove a flywheel at constant speed and was clutched into the hoisting machinery as desired. Such an arrangement ran one way only, to hoist. Lowering the cage or skip was simplicity itself, since the hoist engineer merely threw off clutch and brake, letting the law of gravity function freely. When his indicator showed that the cage was nearing its destination, he eased on the brake to bring everything to a gradual halt. The vertical-cylinder walking-beam arrangement was employed, as has been noted, for the pumping system. Mills themselves preferred the horizontal engine, whose connecting rod ran to a ten-foot flywheel and the belt pulleys or line shafts that ran the mill machinery. Apart from daily oiling and an occasional gland-and-piston repacking, the engine was virtually free of any need for maintenance. Many an engine ran through seven or eight boilers and as many different mill locations before it finally gave up the ghost. Pretentious mills boasted entire gangs of boilers and engines, but the typical one-horse stamp mill used a fifth-hand engine and a tatterdemalion boiler and paid no attention to fuel efficiency as long as the stamps continued to rise and drop.

High-Grading

One mining development which seldom or never appeared in the engineering textbooks but which could make or break a gold-mining operation was high-grading. Like salting, it was so prevalent that the expression for it became common English usage. From the miners' viewpoint, pocketing a bit of extremely rich ore was merely a traditional perquisite of the job. From the owners' point of view, it was outright theft. All gold miners denied doing it; most of them were collectively accused of it; the superintendent hoped at best merely to keep it to the irreducible minimum. Its equivalent in quartz mills was the pilfering by mill hands of rich amalgam from separation and collection machinery. There is no way to estimate the prevalence of, or losses involved in, high-grading. Indeed, if mineowners had been capable of taking exact inventory of in-place ore values and comparing with their returns, they would have had strong leverage in minimizing their losses. Mine superintendents, the men who were in the best position to arrive at estimates of these losses, were not fond of publicizing the figures, since it would constitute a reflection upon their ability to check the habit among the crews. Millmen were somewhat better off. Knowing the tonnage and assay value of incoming ore, they at least could compare it with bullion output (less technical losses, a relative constant) and be able to decide in a hurry whether pilferage was exceeding all bounds of decency.

As long as precious-metal mining was conducted by slaves or convict "lifers" (that is, until the High Middle Ages), there was literally no future in high-grading. This is not to say that engineers, guards, mill technicians, assayers, mint employees, and even casual visitors failed to pocket bits of rich ore or refined metal but only that there was little incentive for the average miner to steal what he could never spend. The moment the miner became a free man, however, his attitude changed remarkably. In this connection Humboldt noted that the Indian slave miner of colonial Mexico was no sooner emancipated than he began to steal—and quite expertly at that.[19]

The economic organization of a mining operation also had a great deal to do with the prevalence of high-grading. As long as miners were joint owners of a mine or placer, they would hardly pilfer from themselves. Mines owned by corporations composed of distant, anonymous shareholders were more often victimized. The worst system of all was

[19] Humboldt, *New Spain*, IV, 246–48.

leasing, whereby the mining corporation rented its mineral rights to an individual contractor, who paid a fixed sum per ton of ore removed or else an equally fixed sum for the right to remove as much ore as he could during an agreed-upon time. The lessee, eager to make a quick profit, was inclined to turn a blind eye toward the practice among his miners, not wishing to antagonize them into slowdown reprisals; a permissive attitude toward high-grading also enabled him to offer lower wages. Additional evils of the leasing system were that it encouraged the lessee to gut the mine of its richer ore, thus severely limiting its economic life, and also to operate the mine in a technically hazardous manner.[20]

The scientific basis of the practice was the geological fact that mineralization in bonanza lodes seldom occurs in uniform distribution. Rather, it is relatively sparse throughout the gangue, or vein rock, for most of its extent, but now and then is highly concentrated in a few favored spots. Thus miners may blast and muck tons of gold ore for days on end without ever seeing any gold at all, even though it is there. Then, without any warning, they may suddenly come upon such a natural concentration. It might be merely a bit of quartz shot through with small flakes or specks of gold. (It is reported that one Cornish miner in early California was nicknamed Old Velvet Thumb in tribute to his ability to detect such a bit of high-grade ore in the darkness of a drift merely by its feel.[21]) Better yet, the crew may find a pocket or small vug which contains an accumulation of native gold up to the size of a marble. Upon such a discovery, it follows that temptation will immediately raise its head.

Since it has already been shown that high-grading was practiced in the Americas long before the opening of the United States mining frontier, no disclaimer of any imputation of its restriction to the Cornish miner need be offered. The Cornish miner was present in notable force on the mineral frontier, however, so perhaps his very numbers in proportion to such other ethnic groups as the Irish account for the fact

[20] Russell R. Elliott, *Nevada's Twentieth-Century Mining Boom: Tonapah, Goldfield, Ely*, 114–15 (hereafter cited as *Nevada's Twentieth-Century Mining Boom*). See also Thomas A. Rickard, "Rich Ore and Its Moral Effects," *Mining and Scientific Press*, Vol. XCVI (June 6, 1908), 774.

[21] Arthur C. Todd, *The Cornish Miner in America: The Story of Cousin Jacks and Their Jennies and Their Contribution to the Mining History of the American West, 1830–1914*, 56 (hereafter cited as *The Cornish Miner in America*).

that a great many stories of high-grading seem to revolve about Cornishmen. Certainly they were no strangers to the practice; Shinn notes that ore stealing was practiced in Cornwall in the seventeenth century and was punished by banishment from the mines and the burning of the offender's house.[22] Arthur C. Todd, who is quite sympathetic to the Cornish miners, notes that the characteristic Cornish lunch pail was for long the favorite receptacle for the removal of high-grade ore, although he is careful to distinguish between mere pilferage and the sequestration of "family ore" by contract miners.[23] In contrast to these distinctions is a tale told to me by Robert Lenon about the Cornish miners of Grass Valley, California, who grew so arrogant in the vice that nearly every one of them had in his basement a water-powered motor, running off the city mains, grinding out high-grade to be amalgamated and sold.

Whatever the miner's ethnic origins, the lunch pail was undoubtedly the first and favorite method for removing his spoils at the end of the shift. A greedy miner would be so heavily laden with an "empty" pail that bandy-leggedness was ascribed to overindulgence in the habit.[24] It was not long before mine superintendents, concerned about their losses, installed functionaries called specimen bosses to examine these receptacles at quitting time. Todd tells about one miner who regularly feigned illness just before time to blast—the traditional final duty of each shift—and gladly accepted being docked in his pay in order to have his pail escape the attention of the specimen boss.[25]

After a time the miners wrought certain improvements upon these crudities. To begin with, raw ore peculiarly enough is frequently recognizable by an experienced shift captain or assayer, who can more often than not name the mine or even the level within that mine whence the specimen has come. In fact, many an attempt to salt one claim with ore of good tenor from another property was instantly defeated after one glance by such an expert at the prospect sample.[26] The same would be equally true in court trials for the theft of high-grade ore. Therefore

[22] Shinn, *Mining Camps*, 29.

[23] Todd, *The Cornish Miner in America*, 56 and n.

[24] Rickard, "Rich Ore and Its Moral Effects," *Mining and Scientific Press*, Vol. XCVI (June, 6, 1908), 775.

[25] Todd, *The Cornish Miner in America*, 56.

[26] Frank A. Crampton, *Deep Enough: A Working Stiff in the Western Mine Camps*, 62 (hereafter cited as *Deep Enough*).

the high-grader wanted to render his accumulations anonymous as swiftly as possible.[27] That was achieved in the home grinder, though it was better done in the mine by dropping the piece of high-grade into a little tube mill—a short length of capped iron pipe with a bolt or piece of rod for a pestle—pulverizing it with a few strokes, and washing out the residue in a coffee cup with any readily obtainable fluid. Such a tube mill was easily concealed behind a loose piece of lagging (very difficult to connect with any given man) and produced gold dust whose provenance was almost impossible to establish.

With their lunch pails subject to search, the miners in turn improved upon their means of concealment and removal of the gold. Humboldt mentions that the Mexican Indians were prone to use for this purpose a capsulelike tube of baked clay which could be hidden about the person even though one was stripped naked for search and required to take the customary dip in a vat of water so as to wash off smears of clay containing appreciable quantities of gold dust. (Humboldt scarcely attempts to conceal his aversion to the Mexican specimen boss's search for these capsules.)[28] The Cornishman or Irishman was required to squat and lift a substantial weight by its handles to the same end. In addition to this time-honored dodge, the American miner used a variety of devices, including a double- or false-crowned hat, capable of holding up to five pounds of high grade; a leg pocket, which was a long sock or cloth tube suspended inside the trouser leg; small pockets sewed inside the waistband of the trousers; and the corset cover, a chemiselike affair of canvas worn beneath the shirt, likewise equipped with pockets.[29] Overweening greed might entail the risk of double hernia, but the risk apparently was accepted quite cheerfully.

The mining companies reacted by adopting an air of benevolent concern for their employees and erecting change houses at the headworks. These were buildings where the miner going off shift could discard and store his soggy work clothes, shower under the benign eye of a supervisor, and don festive attire before walking home through the cold air. The high-graders circumvented this paternalism by organ-

[27] Rickard, "Rich Ore and Its Moral Effects," *Mining and Scientific Press*, Vol. XCVI (June 6, 1908), 774.

[28] Humboldt, *New Spain*, IV, 246–48.

[29] Elliott, *Nevada's Twentieth-Century Mining Boom*, 116.

izing gangs and enlisting the top lander by offering him a share of the proceeds. He could remove high-grade from certain ore cars and cache it in the dump until the opportunity arose for its quiet removal. Failing that, the spoils might be concealed in the sacks which contained timbering wedges or slipped out by way of disused shafts and adits.[30]

The sources contain much less information about pilferage in the quartz mills. It has been noted that the millmen had better information at their disposal, and in knowledge there is security. Nevertheless, there were many points at which the mill was vulnerable to sneak thievery: the ore platform, where incoming ore was received and sledged down to size; the separators, where gold-silver amalgam was freed of pulverized gangue and concentrated for collection; and the retort room, where amalgam was sent to be distilled free of its mercury. Mark Twain relates an incident in which the mill stole a gold ring from *him*.[31] With such lax procedures, it is probable that the underpaid mill hand could get back his own and a bit more to boot. A concrete inference is to be found in the fact that the Consolidated Virginia Mill, of Virginia City, Nevada, had iron covers on its amalgam separators, that these covers were kept securely padlocked, and that only a trusted supervisor was permitted to open them to retrieve their cleanup.[32]

When brought to book in a later day for high-grading, the miner and his unions were quick to point out that the industry's wage scales made the practice almost a necessity. It is quite true that in 1859 a day's wage for the common miner was about $3.50.[33] As late as 1904 a day's wage was still $3.50 in a good many places,[34] even though the cost of living had nearly doubled as a result of the slow expansion of the currency created by the very gold and silver which the mines were hoisting.[35] The mine corporations themselves were faced with rising materials

[30] Rickard, "Rich Ore and Its Moral Effects," *Mining and Scientific Press*, Vol. XCVI (June 6, 1908), 775.

[31] Clemens, *Roughing It*, 255.

[32] Wright, *The Big Bonanza*, 260. Wright calls the separators "settlers," but separators they were.

[33] Lord, *Comstock Mining and Miners*, 61.

[34] Todd, *The Cornish Miner in America*, 219.

[35] According to Mark Sullivan, circulating currency per capita in 1860 was $13.85; by 1892 it was an estimated $20.00 and by 1904, because of the increased production of gold after 1900 by the cyanide process, ought to have been very nearly double its 1859 figure (*Our Times*, I, 158–60, 169).

costs, which, together with the ruthless pressure to declare dividends, caused them to follow the general trend of industry of the period—which was to deny that any remedy was required for a situation which by classical economic theory did not exist. If anything, the mining companies prided themselves on not instituting wage cuts, as had some of the more rapacious captains of industry. Fine-spun economic argument escaped ears deafened by Burleigh drills; the face crews high-graded by night and by day, thus effectively awarding themselves the wage increases (and then some) which the companies refused to grant. Even so, the radical miners' unions found receptive listeners in other industries, such as the Montana copper industry, where high-grading by definition could not exist.

How did the mining camp regard the vice? Deplorably enough, it encouraged it and connived at it. Slugs of melted-down high-grade dust were accepted without question in saloons and honky-tonks.[36] Juries refused to convict accused high-graders. The average juror had good reason to suppose that the next time around he himself might be in the dock and gave as his reason for acquittal the impossibility of proving the origin of the high-grade offered in evidence.[37] Even the clergy, whom the frantic mine companies enlisted in anti–high-grading crusades, dealt very gently with the topic. Crampton tells of one such clergyman in Goldfield, Nevada, who was persuaded by the mining interests to excoriate the sin of theft. The pastor did very well until he came to high-grading. Correctly gauging the temper of his congregation, he concluded his sermon by saying, "But gold belongs to him wot finds it first"—a novel but very welcome exegesis of the Eighth Commandment.[38]

The mining companies, of course, detested high-grading since every half spoonful of gold dust which left the premises in an unauthorized manner represented the loss of ten dollars' clear profit. Mining engineers such as Rickard took a very poor view of the practice, but they could afford the attitude, since their salaries were considerably more than $3.50 a day. Their jobs depended as well upon bullion output, with no excuses accepted, wherefore they were exceedingly quick to deplore

[36] Crampton, *Deep Enough*, 60.
[37] Elliott, *Nevada's Twentieth-Century Mining Boom*, 115.
[38] Crampton, *Deep Enough*, 50.

thievery. On the other hand, the IWW and its ancillary the Western Federation of Miners stated officially that they had no interest in the problem, but their organizers covertly gave aid and comfort to high-grading for both tactical and dialectical reasons.[39]

The issue came to a head at Goldfield, Nevada, in 1907. The camp was a hotbed of high-grading for all the reasons given above: inadequate wages, high-tenor gold ore,[40] and the prevalence of the leasing system.[41] Discovering that it was losing up to 40 per cent of its bullion, the Goldfield Consolidated Mining Company first announced that it intended to institute private detective patrols of the working chambers. The miners declined to permit that. The Goldfield Consolidated then instituted change houses. Urged on by the IWW, the miners threatened to tear up the camp by the roots. At length United States troops commanded by Major Frederick W. Funston tranquilized the community and, incidentally, confirmed the change-house rule.[42] Since Goldfield's mines went into permanent *borrasca* soon afterward, it is difficult to determine how much high-grading suffered from those measures.

Apart from putting it into circulation as the currency of the country, how did the miner dispose of his high-grade? A prime fence seems to have been the seedy assay house. The operator discounted high-grade that he received and concentrated it by chlorination in homemade barrel reactors; one thrifty fence even sold his process tailing for use in salting prospects.[43] Elliott says that Goldfield had over fifty such bogus houses during the heyday of high-grading. Morris says that the camp had "hundreds" of the houses, not one in three doing an honest

[39] Elliott, *Nevada's Twentieth-Century Mining Boom*, 114–15; Morris, *Desert Gold*, 16–17.

[40] Rickard stated that in the Frances Mohawk lease there was an eight-inch ore ledge which went $250,000 to the ton ("Rich Ore and Its Moral Effects," *Mining and Scientific Press*, Vol. XCVI [June 6, 1908], 774).

[41] Common causes have common effects. About 1864, Henry Wickenburg's famed Vulture Mine in Arizona was nearly high-graded to death, losing 20 to 40 per cent of its ore to the "forty thieves" who labored there. Unlike the Goldfield operations, which were based on fixed-time contracts, Wickenburg sold his ore "in place" at fifteen dollars a ton, but the consequences were the same (Dunning and Peplow, *Rock to Riches*, 62–63).

[42] Morris, *Desert Gold*, 17; Elliott, *Nevada's Twentieth-Century Mining Boom*, 116–25.

[43] Rickard, "Rich Ore and Its Moral Effects," *Mining and Scientific Press*, Vol. XCVI (June 6, 1908), 774–75.

business or even inviting it.[44] Crampton, who was running such an assay house at the place and time, admitted buying gold ore on the strength of the seller's declaration that it came from a mine "back in the hills" but asserts that he drew the line at refining obviously stolen dust and ore offered by honky-tonk owners or by the business managers of the many prostitutes who frequented the camp.[45] The spurious assayers then sold the bullion directly to the United States Mint at Denver, which apparently asked no questions.[46]

Another cover story—and from the historian's viewpoint a more pernicious one—was for a buyer to pretend possession of a small but very high tenor "secret" mine. Periodically he would disappear into the wilderness with elaborate precautions, then pop up a few days later with a hundredweight of "jewelry" ore and gold dust—each piece of which had been high-graded from someone else's mine. Few people and no owners were deceived, but it was virtually impossible to prove otherwise before a jury composed of grinning Cornishmen. Crampton stated flatly that Walter Scott ("Death Valley Scotty") obtained his great wealth not from a secret mine but from large-scale trafficking in high-grade at Goldfield.[47] Similarly, the Lost Dutchman Mine of Jacob Walz, which supposedly exists somewhere in the Superstition Mountains east of Phoenix, Arizona, is reputed to have no better foundation in fact than Walz's labored explanation of his possession of ore which looked remarkably as though it had come from the Vulture Mine.[48]

Courtroom Mining and Process Peddling

A legalized form of thievery had its historical roots in the consensus, erected into common law by the miners' meetings, that the discoverer of any part of a mineral vein, lead, lode, or ledge should be vested with title to all of it, though it might "throw" as far west as China or dip as far down as hell.[49] The game lay in locating, or claiming, the lode where

[44] Morris, Desert Gold, 16–17; Elliott, *Nevada's Twentieth-Century Mining Boom*, 115–16.

[45] Crampton, *Deep Enough*, 60.

[46] Rickard, "Rich Ore and Its Moral Effects," *Mining and Scientific Press*, Vol. XCVI (June 6, 1908), 774.

[47] Crampton, *Deep Enough*, 52–53.

[48] Duning and Peplow, *Rock to Riches*, 63.

[49] This law was changed by a United States statute in 1866. The new law, drawn by Nevada's Senator William M. Stewart, provided in effect that extralateral

it cropped out of the ground (at its apex), since he who held the apex held all. This was a simple and logical rule, provided that mineral veins were always continuous and discrete—which in fact they seldom were. Because of this geological difficulty, and because only the Almighty knew for certain the exact extent and ramifications of an undeveloped lode, such uninspired reasoning soon revived an ancient and sometimes highly profitable occupation—courtroom mining, or, as it was more formally called, apex litigation.[50] These law suits ranged from the serious to the frivolous and the motives behind them from genuine damage to mere speculative blackmail. Observant contemporaries noted that the amount of litigation seemed to vary directly with the profits of a given mine and quickly drew appropriate conclusions. Indeed, apex litigation was exactly like modern medical malpractice litigation, in that the facts were so hard to determine that sheer legal showmanship made a great deal of difference in the judgments and the awards.

A famous case in point was *Ophir Mining Co.* v. *Burning Moscow Mining Co.* (1863), in which arrant pilferage may have been involved. The claim boundaries of the two companies adjoined, but whereas Ophir lay upon the main Comstock outcrop and ore body, Burning Moscow sat upon ground which had no surface apex at all. Undaunted, the Muscovites merely put down a shaft until they hit the westward-dipping lode, then drifted over the boundary and began helping themselves to ore which Ophir considered its own. Pitched battles took place below ground as well as in the Nevada courts. Burning Moscow offered the plea that it was but following a subterranean apex of its very own

---

rights extended only across the side lines of lode claims; the end lines were to be vertical and marked as such. If a square or diamond-shaped claim existed, the locator must specify which opposite sides were the side lines for just this reason; and if a triangular *hueco* was left, one must stake it not as a triangle but as a keystone-shaped figure, even if the short side was but one foot long. This short side and its mate then became the end lines (Robert Lenon).

[50] Apex litigation was probably as old as the hills. Lord quotes the Spanish jurisconsult Sáenz as referring to some apex litigants as "pretended miners" and quotes Bauer to the effect that "these rascals should be banished from the mines as pilferers" (*Comstock Mining and Miners,* 116). In all fairness, however, Lord notes that lode locators "were in no hurry to mark [their claims], . . . as every man naturally wished to cut off the richest slice of the prospective bonanza and was not disposed to cut the loaf until he knew its contents," whereupon he would sidle or "float" his markers in the most promising direction (*ibid.,* 49–50).

under the common law of lode mining, whereas Ophir had been located under the provisions of placer common law and hence was entitled only to what it found on the surface.

Had Ophir not objected loudly and soon, Burning Moscow might well have had its way, but counsel for the plaintiff, William M. Stewart, went back into the exiguous records of Gold Hill to show that the original locators (McLaughlin, O'Riley, Comstock, and Penrod) had clearly staked Ophir as a lode claim. Stewart was greatly influenced by his experiences in this and similar cases, and when he went to the United States Senate he virtually dictated the new United States mining code, which was calculated to reduce such piracy.[51] He was probably also much embarrassed when the ore ledge later reversed its dip, and he had to represent Ophir with a brief whose arguments exactly reversed his successful pleading in the suit against Burning Moscow.

At its rock-bottom worst, apex litigation began when a would-be plaintiff had, or arranged to obtain, title to property adjacent to that of a prosperous mine. He next hired a geologist to "discover" that a shoot of the high-grade lode cropped out on his own property and to theorize somehow that the tail of the shoot wagged the dog of the lode. The next step was to retain a high-powered attorney, one who specialized in apex litigation on a contingency-fee basis, to bring suit against the prosperous mine. The attorney then filed a brief on the law side[52] of the appropriate court, swearing that the plaintiff either had priority of discovery (thus entitling him to the whole) or at worst was entitled to receive lucrative remedies for the damages he was suffering from the depredations of a soulless corporation. The hope, of course, was that the defendant mining company would settle out of court for a substantial sum, irrespective of the facts, in order to be rid of a dangerous nuisance. This extortion might be speeded up by persuading a com-

[51] Courtenay de Kalb, "William Morris Stewart," *Mining and Scientific Press*, Vol. XCVIII (May 1, 1909), 599–600.

[52] Apex cases were invariably brought on the law rather than the equity side of the courts to ensure that the facts would be decided by a jury instead of a judge, or chancellor in equity, alone. Juries were far more easily swayed by scenery-chewing theatrics than were aged and cynical wearers of the judicial ermine. Needless to say, all efforts to reform the system were bitterly and successfully resisted by the fraternity of contingency-fee pleaders and professional experts, who wanted no part of the proposal to have the cases heard in equity or decided by a technically qualified referee (Rickard, *Retrospect: An Autobiography*, 67–68).

pliant judge to enjoin the defendant mining company from removing any more ore until the suit had been completely adjudicated.

If the defendant refused to be blackmailed, the plaintiff reinforced his argument by importing at enormous expense a clutch of professional mining experts to testify to the validity of his complaint and to affirm the hypotheses of his tame geologist. He invariably hired as many of these gentlemen as he could afford, on the theory that numbers were more impressive to a lay jury than would be the technical facts. The defendants counterattacked by bringing in an equal or larger number of geologists and experts who, like those of the plaintiff, were handsomely paid for interpreting the facts the way their clients wished. Then juries would be treated to the spectacle of two teams of nationally renowned experts solemnly contradicting each other, each using technicalities which only a specialist could understand.[53] About the only relief from the flood of words which hard-pressed judges and juries could hope for was the prospect that a cunning attorney would trap an expert in some inconsistency, then systematically reduce him to tears and screaming upon the witness stand. Otherwise, there was little for the jury to do but retire and flip a coin, literally or figuratively—this being the basic fact which plaintiffs were prone to rely on, plus the human reaction of sympathy for the oppressed individual playing David to the corporate Goliath. In short, apex litigation was a sorry business but so profitable withal that many an expert spent his entire career and amassed a fortune in fees, doing nothing but hiring out his reputation to litigants.

Another nuisance of more excusable motivation was found in the prevalence of "process peddlers," whose reason for being was the undeniable fact that millions of dollars, particularly in silver values, were getting away from the mills in use during the period. Wright tells about this:

> The early millmen knew but little about working silver ores, and all manner of experiments were tried. . . . A more promiscuous collection of strange drugs and vegetable concoctions never before was used for any purpose. The amalgamating-pans in the mills surpassed the caldron of Macbeth's witches in the variety and villainousness of their contents. Not content with bluestone [sulphate of copper], salt, and one or two other simple articles of known efficacy, they poured into their pans all manner of

[53] *Ibid.*, 66–67.

acids; dumped in potash, borax, saltpeter, alum, and all else that could be found at the drug-stores, then went to the hills and started in on the vegetable kingdom. They peeled bark off the cedar trees, boiled it down until they had obtained a strong tea, and then poured it into the pans, where it would have an opportunity of attacking the silver stubbornly remaining in the rocky parts of the ore. The native sagebrush, which everywhere covered the hills, being the bitterest, most unsavory, and nauseating shrub to be found in any part of the world, it was not long before a genius in charge of a mill conceived the idea of making a tea of this and putting it into his pans. Soon the wonders performed by the "sage-brush process," as it was called, were being heralded through the land. The superintendent of every mill had his secret process of working the silver ore. Often when it was supposed that one of the superintendents had made a grand discovery, the workmen of the mill were bribed to make known the secret. To guard as much as possible against this, the superintendent generally had a private room in which he made his vile compounds. "Process-peddlers," with little vials of chemicals in their vest pockets, went from mill to mill to show what they could do and would do, provided they received from five thousand to twenty thousand dollars for their secret. The object with many inventors of "processes" appeared to be to physic the silver out of the rock, or at least to make it so sick that it would be obliged to loose its hold upon its matrix, and come out and be caught by the quicksilver lying in wait for it in the bottom of the pans . . . . The "process-peddlers" finally became a worse nuisance than even lightning-rod men have been, the limited space of the country to which they were confined being considered.[54]

Flotation

Some good was almost bound to come from all this experimentation, even if only by accident. One example is chlorination, which will be dealt with later. In another instance the lead came in the observation that metallic sulphides had a tendency to stick to and float off with greases and oils, whereas silica particles had no such inclination. Since even the slightest contamination with oil prevented effective panning or amalgamation, millmen had learned to keep their crushers and stamps and prospectors their pans scrupulously free of oil and grease drippings, but it was another matter to turn the phenomenon around and put it to constructive use. In the diamond diggings of Africa and Brazil workers dribbled the blue earth, or placer sands, down the surfaces of

[54] Wright, *The Big Bonanza*, 92–93.

grease-covered tables, then hand-picked the entrapped diamonds, but this procedure obviously would not do for the dustlike sulphides. Some experimented with heavy oils poured into mill pulps but found that prohibitively large quantities of oil were required and that yields did not come up to those of conventional methods.

There is a charming tale—unfortunately untrue—that oil flotation was first discovered by a woman, Carrie J. Everson, who lived in the Denver area. The story goes that Carrie was hand-laundering either ore sample bags or work overalls soiled by both sulphides and oil and that her attention became focused upon the black froth which rode up on her suds. Upon investigation she found the froth to be composed of metallic sulphides, whereupon she patented her process and presumably made a mint. Alas for romance! Upon investigation it appears that flotation was gradually worked out by unglamorous engineers and that Miss Everson must be relegated to that apocryphal sorority which includes Peggy O'Neill, Nan Britton, Anna E. Carroll, and others who by one means or another supposedly influenced the menfolk to influence American history.[55]

Flotation begins by mixing ore with a small quantity of light oil right in the ball mill, then introducing it to a water bath which contains an even smaller amount of acid (plus a touch of sodium cyanide or other reagent). The acid is the key to the process; it etches the mineral crystals, giving them a clean surface which enables the oil to adhere tightly.[56] The oil in turn greatly assists the formation of bubbles with a high surface tension, which are created by a stream of low-pressure air entering the bath from a sparger below. The air creates bubble; bubble is enclosed in monomolecular oil film; sulphide crystal adheres to oil; and all sail up in a froth to the top of the bath. Here the froth drools over the side of the flotation cell into a launder (trough), which carries it away, while the pulverized gangue is washed away in the current of

[55] Rickard disposes for good of the Carrie Everson legend by tracing carefully the true origins of flotation, step by step. Its first successful use did not come until the twentieth century, and then in New Zealand (*A History of American Mining*, 397ff.).

[56] It was noticed in development that minerals of metallic luster would be susceptible to flotation but that free gold coated with a manganese film would not. This problem was solved by chemical etching; matte-surfaced white-lead carbonate will not float, but a bit of sodium sulphide in the water cleans its surface, and up it pops (Robert Lenon).

the bath, coming at last to rest on the tailing heap. As with any laundry work, mild heat, gentle agitation, and soft water greatly assist the process. At the end of the line, the sulphide concentrates are vacuum-dried and bagged or shipped in carload bulk to the smelter.

Flotation was not widely used until about 1910, and in the precious-metals field it had already been displaced by cyanidation. It was the salvation of the copper industry, however, as the end of the high-grade, direct-smelting ores of Jerome and Butte came in sight. The copper states still had a virtual infinity of copper left in the huge dissemination deposits of the great desert stocks, but the engineers had been at a loss to deal with ores which ran little better than 2 per cent copper at best. They provisionally worked out a chemical leaching process which involved steeping the ores in a weak solution of sulphuric acid. Presently out of the heap would trickle a stream of bluish copper-sulphate solution, which could be exposed to scrap iron (actually shredded and detinned "tin" cans) to get metallic copper. The process worked just well enough to be tempting, but not well enough to be satisfactory. Moreover, it failed to extract the tiny but valuable salting of precious metals invariably present. Flotation overcame almost every shortcoming of the leaching process, which was forthwith relegated to scavenging values from abandoned workings and waste dumps.[57]

[57] In Arizona at present copper-sulphate leaching produces 30 million pounds of copper annually and is likely to produce more as it is applied to copper-silicate ores which have proved resistant to economical working by any other method, including flotation. The sulphuric acid is a by-product of smelter gases, so that one industrial hand washes the other, so to speak. Sulphate leaching, by the way, is not suitable for reworking tailing heaps, which are comparatively impervious to water. The "pregnant liquor" is handled in redwood conduits, since cement dissolves in its presence and iron is, for obvious reasons, unsuitable.

CHAPTER IX

The Comstock Lode

SOME MILES EAST of Lake Tahoe in the central Sierra Nevadas the eastern edge of the mountains descends in an abrupt escarpment to the low-lying flats of the Great Basin. Here the snow-fed Carson River waters a pleasant valley which lies between the Sierras to its west and the lower, hogback mountain chains so characteristic of the desert to the east. Since the passes of the Overland Trail were somewhat north of Carson Valley, it was little visited until May 15, 1850, when a party of Latter-Day Saints entered it, seeking pasturage and water during the course of their journey to California. There they camped, waiting for the Sierra snows to melt. One of the party, a young man named William Prouse, said to have passed that way before and to have had some idea of what he would find, began to pan along the banks of a small stream which emerged from a cleft in the desert mountain chain on the northeast and which flowed into the Carson River near the camp site. Prouse found some colors in his pan and persuaded a few friends to prospect with him farther up the stream. They panned their way into the mountain gulch which they named Gold Canyon, followed its curving course upward a mile or so, and came to a cleft at which it abruptly narrowed. They named the cleft the Devil's Gate, and at that point Prouse and a companion, Nicholas Kelly, found some gold nuggets in fissures of the colorful country rock. That cleft and that rock are still there, accommodating the access highway, and the innumerable threadlets of visible quartz are enough to set a hard-rock prospector's ears twitching.

The Mormon party presently went its way, but word of Prouse's discovery must have been bruited about, for in August of the same year a certain Don Ignacio Paredes and a following of peons were observed dry-washing the sands in Gold Canyon. The cleanup apparently was not profitable enough to detain Paredes, but during the course of the next year (1851), a small colony of wandering placer men began to accumulate about the mouth of the gulch in a settlement they called Johntown. More Mormon settlers came to farm the Carson Valley and sell provisions to the placer camp, as was their profitable custom. Sometimes relations between the two groups were good, and at other times they were not. The population of Johntown enlarged when winter rains and spring runoff provided water for placering, then diminished during the dry summer and autumn months, but the camp survived and even obtained a few of the amenities of life, including a saloon. Johntown lasted for seven years, collectively grossing perhaps eighty thousand dollars annually in cleanup dust, but even so made little more than grub. Indeed, by 1857 it was thought that the sands had been riffled out and that Johntown would soon be a ghost camp.[1]

At that point two young brothers, Ethan Allen and Hosea Ballou Grosh, came to the camp, being convinced almost to the point of fanaticism that there was silver in the hills about them. They followed the creek to its only fork, the American, and thereabouts hit what Allen Grosh felt sure was an enormous ledge of silver mineralization. By August 16, 1857, two different field assays had convinced them that they were onto something very good.[2] Three days later, disaster struck. Hosea Grosh hit himself in the foot with a pick and in a fortnight was dead of gangrene. Allen Grosh delayed overlong settling his affairs, for when he and a companion attempted to cross the Sierras to California, they encountered severe snows, and in December, Allen Grosh died of starvation and exposure. The Grosh brothers had always been dogged by bad luck; even their prospective silver bonanza ultimately turned out to be only a low-grade ore shoot which was never profitable to work.

By 1858, the remaining inhabitants of Johntown were doing more

[1] Contrary to supposition, the placer man could not merely follow the stream and color to find the Comstock Lode, for the color did not come from the lode at all but from small, low-grade gold shoots which cropped out in the side branches of the canyon. The Comstock itself was so geologically recent that insufficient time had elapsed for the placers below it to become very rich. Today Gold Canyon is dotted with small mines and mills, but they date from cyanide times (after 1900) and not from the earlier period.

[2] One test was by "burning with damp gunpowder"; the other, by "cupellation." William Wright participated in an archaeological dig in 1860 which uncovered crude furnaces believed to have been used by the Grosch brothers (*The Big Bonanza*, 14–15).

rambling and game hunting than digging. During the course of a trip up the creek, James ("Old Virginia") Finney discovered a mildly attractive outcropping stained with yellow limonite. Knowing that limonite blossom was sometimes an indication of gold, Finney persuaded some of his fellows to go in with him and locate it as an association placer claim. Moving up to the outcrop, they called their new camp Gold Hill. It was just across a low watershed from the slopes of the tallest peak in the district. This mountain they promptly christened Sun Peak, undoubtedly because its summit was their alarm clock, it being the first point to catch the light of the rising sun as it appeared above the lower range lying east of them. Finney and his associates worked the yellow outcrop by amalgamation in an arrastra, cleaning up as much as twenty-five dollars a day, not a bad return, although water for working became scarcer as returns became richer.

Other prospectors drifted in, scenting a going concern. Finding Gold Canyon pre-empted, they went around Sun Peak and began to work up Six-Mile Canyon, another gulch two or three miles north of Gold Hill. This canyon ran straight up Sun Peak itself, its head washes fading out about halfway up the steep slope. These newcomers found color in their rockers, but it was so pallid that buyers would not accept the "bogus gold" until it had been quietly adulterated with California placer dust. To add to their difficulties, the soil of Six-Mile Canyon was clay, which made puddling necessary. Thus it was that in May, 1859, two sons of Erin, Patrick McLaughlin and Peter O'Riley, were sweating at the head of Six-Mile Canyon and discussing abandoning their present location in order to join the current Mono Lake rush to the south. They lacked a stake for the move, however, and so resolved to make one last try at obtaining something worthwhile from their claim.

More placering demanded more water, and McLaughlin and O'Riley had depended upon a slight trickle which ran down the gulch from the slopes above. To improve matters, they crossed the low saddle between the head of the canyon and the side of Sun Peak and began to excavate a reservoir to accumulate more water from the rill. Unbeknownst to them, they had left the surface of the country-rock stratum and were now working the point where its upturned lower edge touched the entirely different rock of the mountain above. On June 8, as they commenced to dig, their shovels cast up an unattractive mixture of

Fig. 69. Henry T. P. ("Pancake") Comstock. From William Wright [Dan deQuille], *The Big Bonanza* (1964).

yellowish sand mixed with white quartz and an unidentifiable black, friable mineral. They experimentally washed a peck of the mixture, recovering a substantial amount of pale yellow metal—undeniably gold, but in their opinion heavily contaminated with some base metal whose nature they lacked sufficient art to identify. The quantity of the color was more than encouraging, however, and the pair made ready to enlarge their operations.

At that moment an unappetizing local character, Henry T. P. ("Pancake") Comstock, happened along (Fig. 69). It is not likely that McLaughlin and O'Riley greeted him with enthusiasm, for Comstock was preceded by a reputation as a blowhard whose life's ambition was to evade the curse of Adam at the expense of others. Old Pancake cocked an eye into the rocker and, taking opportunity by the forelock, mendaciously informed his hosts that they were trespassing upon a "ranch" owned by himself and his partner, Emanuel ("Manny") Penrod. There is little doubt that Comstock had no claim whatsoever to the land and only a somewhat nebulous equity in the water; yet McLaughlin and O'Riley gave in meekly, agreeing to assign Comstock and Penrod each one hundred feet of their prospective location. Comstock left them to their labors, and the two Irishmen worked on a few more days before realizing they were engaged in lode mining in a placer district. That would never do. They concealed their prospect hole, then crossed over to Gold Hill to agitate for a miners' meeting to enact a lode-claim law.

The residents of Gold Hill, who had themselves got down to hard rock, agreed to convene the meeting on the evening of June 11. The preamble of their code is sufficiently significant to be quoted in full:

> Whereas, the isolated position we occupy, far from all legal tribunals, and cut off from those fountains of justice which every American citizen should enjoy, renders it necessary that we organize in body politic for our mutual protection against the lawless, and for meting out justice between man and man; therefore, we citizens of Gold Hill do hereby agree to adopt the following rules and laws for our government.

It was an excellent statement of the basic principle of the common law, but it can be surmised that the Gold Hill version was not the original product of some anonymous Blackstone, but rather was borrowed verbatim from California usage of long standing, originally worked out by lawyers in the Sacramento camps.

On the following day Comstock, Penrod, O'Riley, and McLaughlin staked their "Ophir" outcrop at the saddle above Six-Mile Canyon, each taking three hundred feet, plus the traditional claim for discovery. Other parties took up adjoining locations, so that there grew up two different districts, separated by about a mile—the northern "Comstock" set of claims, and the southern "Gold Hill" group, revolving about Finney's discovery.[3] There matters momentarily rested, with all parties

[3] The Gold Hill locations were the basis for the Best & Belcher, Crown Point, and Yellow Jacket mines (among others), while the Comstock sector held the

convinced that they had found gold lodes of moderate tenor, a conclusion readily reached in view of the existence of low-grade gold shoots down in the throat of Gold Canyon. Rockers were discarded in favor of quartz picks, and arrastras were constructed from the handy gray diorite[4] of the mountain (ultimately renamed Mount Davidson in honor of a San Francisco banker who helped incorporate the Ophir Mine).

The bluish black mineral gave continual trouble in its refusal to be separated from the scanty gold values. The placers of Gold Canyon had held the customary and expected brown-black hematite sand and slickers, but the bluish tinge of the new sand was unfamiliar to the miners. It clogged riffle bars, could not be air-winnowed, and generally made a nuisance of itself. Someone was finally impelled to send a specimen to the assay house of Melville Attwood in Placerville (formerly Hangtown), California. Atwood returned his report on June 27, 1859, stating that the yellow sand was porphyry (syenite) heavily mineralized with silver chloride, and that the black sand was silver sulphurets. Taking one thing with another, the sample went $3,000 in silver and $867 in gold to the ton! The California newspapers got wind of the bonanza, published it in crack-of-doom type, and the rush to the Comstock Lode was on.[5]

The original partners in the various Gold Hill and Comstock claims[6] were promptly bought out by San Francisco mining promoters. As suggested earlier, prospectors did not understand or enjoy mining, had little liking for a sedentary life, and certainly were in most cases lacking the promotional talent necessary to organize, finance, and defend a

Ophir, California, Mexican, Gould & Curry, Savage, Hale & Norcross, Chollar-Potosí, and, ultimately, the Consolidated Virginia.

[4] For the sake of clarity, some distinctions should be made. Dolerite is a basaltlike lava. Dolomite is a sedimentary rock composed of a mixture of magnesium carbonate and limestone. Diorite is a gray igneous hornblende-feldspar resembling granite. Contemporaries continually referred to the diorite as "granite," and to the valley syenite as "porphyry," even though syenite is diabase, lacking the high silica content of true porphyry.

[5] The chief source on discovery of the Comstock Lode is Lord, *Comstock Mining and Miners*, 11–55; and additional material has been abstracted from Wright, *The Big Bonanza*, *passim*, and Shinn, *The Story of the Mine*, *passim*. Corrections to certain mistaken conclusions of these three writers are to be found in Smith, *The History of the Comstock Lode, 1850–1920*, 1–11.

[6] The Gold Hill distinction will not be used after this point, all the sulphuret ledges on the flank of Mount Davidson being called the Comstock Lode.

first-class venture. The partners sold out for seemingly paltry sums—O'Riley received about forty thousand dollars for his share of the Ophir Association claim.

Evidently the sellers were satisfied with their price, and the purchasers risked not only that amount but in addition the even more risky costs (which could be considerable) of development. For every fortune made in mining promotion another fortune was spent ultimately to provide palatial lodgings for gophers. What the prospectors who sold out did with their money was their business, but it is well to note that few, if any, demanded or bought shares in their own discoveries, even though they were in good position to pay their stock assessments.[7] A poll of the original Comstock claimants made seventeen years subsequently revealed that half of them were dead, most of the remainder were living in reduced circumstances, and none were rich. They had come to this pass by being "garrulous, imprudent, and spendthrift," although a certain susceptibility to feminine wiles had (as it is phrased delicately) "disastrous consequences in many cases."[8] Old Pancake Comstock was particularly notorious for the latter failing, although, in the opinions of his contemporaries, he had sufficient other shortcomings for three men.

This book will not attempt to deal with the logistics of creating and supplying a community of fifteen thousand souls, two dozen major mines, as many quartz mills, and hundreds of speculative mines, located in a desert two hundred miles from its supply base and moreover separated from that base by one of the most imposing mountain ranges in the world. The details are fully recorded elsewhere and will be adverted to only as necessity impels in the course of dealing with developments in Comstock mining and milling. A representative set of statistics may be afforded, however. In the first year of major operations (1859) the Ophir ledge was worked in open-pit fashion, the ore up-

[7] Mining stock was usually sold in "feet," or shares of the whole, equivalent to the "lay" of a whaling venture. When development money was needed, the board assessed each foot a pro rata sum. Failure to pay one's assessment meant automatic reversion of the share to the corporation. Thus a new mine issue was cheap to buy into but expensive to keep. Clemens noted that some mine boards in Virginia City sold feet in worthless holes and lived on the assessments they levied upon the absentee shareholders (*Roughing It*, 250).

[8] Smith, *The History of the Comstock Lode, 1850–1920*, 11–12.

graded by hand or concentrated by arrastra grinding and amalgamation, and the output sent to San Francisco by ore wagon, railroad, and riverboat in turn. The Ophir received three thousand dollars a ton, but paid close to one thousand dollars a ton in transportation and refining costs. Costs of mining and concentration at Virginia City are not recorded, but must have amounted to another very substantial fraction of the whole. The Spaniards were right when they said that it took a gold mine to "run" a silver lode.[9]

Some efforts were made to smelt the Comstock ore. It was sufficiently lead-rich to permit reduction, but the wood in the vicinity was in too great demand for construction and mine timbering to permit burning it for charcoal. To import coal or coke was no less expensive than to ship out concentrates. As a result, smelting never caught on in the Comstock. Such milling as was adopted was confined to other techniques.

By 1860 open-pit mining had ended, and it was time to delve underground. That made necessary some thought about the fundamental geology of the lode. By then it was being theorized that Mount Davidson was an enormous cone of barren diorite which had quite recently been forced upward through the pre-existing strata. The edges of these strata had in turn been metamorphosed, bent, and tilted upward during the uplift process, thus becoming fractured and brecciated to a remarkable degree. This country rock was the unsightly pale-yellow syenite; the zone of its uplift and fracture where it lay in contact with the Mount Davidson diorite was also the zone of mineralization, so that the silver-quartz lodes were sandwiched between the foot wall of the mountain and the hanging wall of the crazed porphyritic ground on the eastern slopes of the district. Now and then large detached blocks of this syenite were found embedded in the silver ledges. Miners called such barren inclusions horses and were not delighted to encounter them (Fig. 70).

The silver ledges themselves were very high grade. In the first mining operations they were found dipping west toward the mountain and widening rapidly with increasing depth. It was also determined that the major deposits were located upon an arc whose focal point was the summit of Mount Davidson. This arc appeared to be broken off abruptly

[9] *Ibid.*, 15–16.

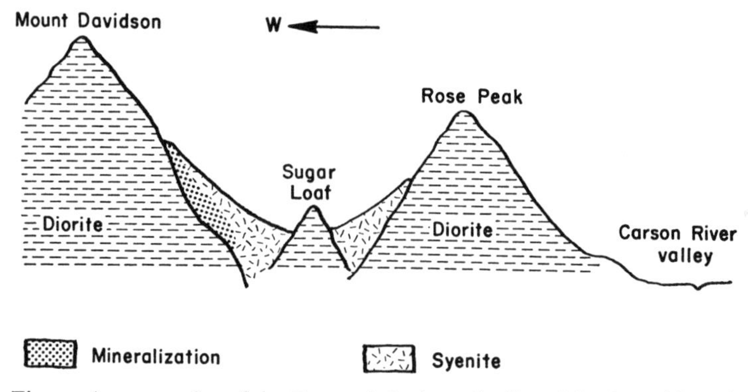

Fig. 70. A cross section of the Comstock Lode on the line of the Sutro Tunnel, looking north (not to scale).

by *borrasca* ground at its northern extremity. Virginia City grew up virtually atop the Comstock mines, as did the settlement of Gold Hill upon its own group, south of the low divide. Gold Canyon afforded and still affords a transportation route from the Carson River Valley, which went around the flank of the Rose Peak range, lying directly across the valley to the east, and between Virginia City and the Carson River. The old creek bed in Gold Canyon constituted a convenient natural flume for mill-tailing disposal, and one of the first mills in the district, the Rhode Island, was located a stone's throw above the Devil's Gate, where its foundations may still be observed in place (Fig. 71).

The Ophir Mine

Since the Ophir Mine was in many respects the prototype mine of the district, an account of its technical and legal adventures may serve as a summary for all. When the sides of its open pit began slumping, the Ophir opened an incline to follow the western dip of the lode. At the 180-foot level the ore body widened to 45 feet. It was becoming apparent that the lode was not so much a vein as a massive deposit standing on its face. Its sulphuret quartz was so friable that little blasting was required; the miners could hack it out quite readily with quartz picks. By no stretch of the imagination, however, could such ore or the syenitic hanging wall be considered self-supporting, nor could the conventional tunnel-set system of support be used. Timbers of the length required

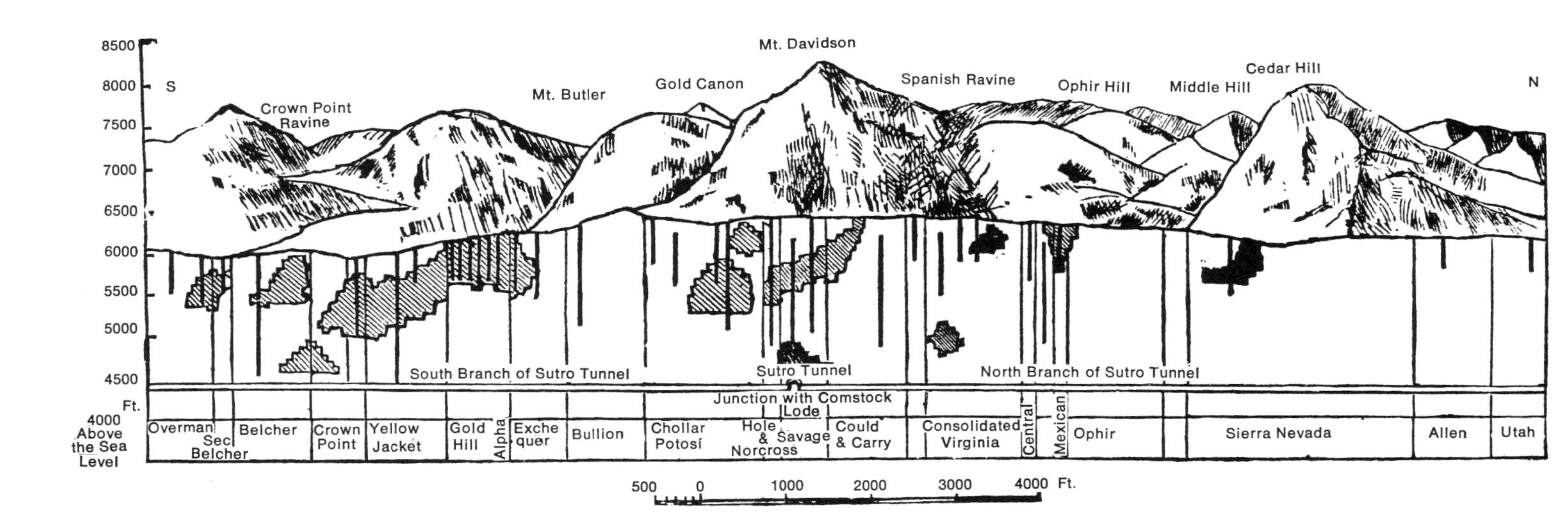

Fig. 71. A section across the Comstock Lode on the line of the Sutro Tunnel north-south drainways, looking west. From J. Bernard Mannix, *Mines and Their Story* (1913).

were unobtainable in nature, and when extralong posts were spliced together, they buckled under the stress. Room-and-pillar support was equally out of the question, since neither the ore nor the country rock had any structural strength to speak of.[10] What was required was support of extraordinary strength and rigidity in three dimensions, fashioned economically of available materials and capable of extension in *any* combination of directions as the ore body was stoped out.

At this juncture Philipp Deidesheimer, graduate of the Freiburg School of Mines and a distinguished mining consultant in California, was summoned to diagnose and prescribe.[11] He devised a square-set, or laid-over, cube whose edges appear to have consisted of eighteen-by-eighteen-inch timbers,[12] mortised together and capable, within reason, of supporting the heavy ground of the mine as well as dozens of its fellows in the great stopes (Fig. 72). The square sets made mining history, being considerably more economical than cribbing.[13] Even so, fortunes were required to fell the trees in the high Sierras, transport them by flume for hundreds of miles, then square, frame, and install them in the mines. Even when selected columns of square sets were reinforced with rock fill, they served their purpose only when the ground was quiescent. When ground wants to move, nothing whatsoever can restrain it, and the Comstock was restless. The square sets were extremely vulnerable to fire in the presence of the stack draft created by hundreds of feet of rising air after ventilation was improved. Almost as bad was the dry rot which flourished in the damp heat of the lode. The square sets began to rot insidiously almost as soon as they were installed. Despite these drawbacks, they worked, especially when the miners learned to keep them gobbed up with waste as tight against the working

[10] Wright, *The Big Bonanza*, 90.

[11] In its day the great Freiburg School of Mines was the world's leading institution of scientific mining technology. The school was originally founded by reason of its proximity to the silver and lead mines of the Harz Mountains. Dozens of its graduates came to the United States in the 1860's and 1870's.

[12] Contemporary newspaper accounts state that the timbers were eighteen by eighteen inches. Magazine illustrations, however, show miners standing beside timbers which, unless the miners were midgets, must have been at least thirty by thirty inches. Perhaps these massive frames were used at hoist shafts or winzes, to help keep them in plumb (Robert Lenon).

[13] Cribbing consists of squared timbers piled up, log-cabin–fashion, for the support of heavy ground.

Fig. 72. Square-set timbering. From William Wright [Dan deQuille], *The Big Bonanza* (1964).

face as practicable, so as to minimize the wobble that is inherent in any geometrical form but the triangle (Fig. 73). From the moment of his invention, Deidesheimer was regarded as the presiding genius of the Comstock, and his blessing was eagerly sought—though not always obtained—on every technical decision of moment. Ironically, Deidesheimer later went through bankruptcy as the result of injudicious mining investments.

Hardly had the support problem been resolved when another difficulty became prominent. The ore bodies were full of water, some of

Fig. 73. Square-set timbering. Reproduced by permission of Buck O'Donnell and Shaft and Development Machines, Inc.

which had accumulated from surface seepage over the millenia, but most of which was of plutonic origin. This water was scalding (the maximum temperature recorded was 170° F.), and had evidently percolated up from profound depths, both carrying and depositing the sulphurets of the lode itself. The theory was considered proved when, with concurrent massive dewatering at a later date, the quite-distant Steamboat thermal spring, a landmark of the Overland Trail, ceased to flow. The engineers sighed at the prospect of working stopes awash with hot water and reeking of sulphides in solution. In early phases of operation, when the mines were being worked from shafts, small Cornish pumps were installed; Deidesheimer was thought extravagant when in 1861 he recommended that the Ophir install a forty-five-horse-power eight-inch pump. To assist the pumps, horizontal drainage and haulage adits were driven into the mines a few hundred feet down the slope (four or five city blocks in steeply terraced Virginia City).[14] In this way the lift of the water could be minimized, the strain on hoisting and pumping machinery eased, and ventilation greatly improved.

To judge from contemporary accounts, the ore body was presently found to change its dip from west to the vertical. It was now worked by first driving an ample shaft downward until the wall rock of Mount Davidson was reached. After the shaft was timbered, drifts were extended from it laterally at about one-hundred-foot vertical intervals, the men therein picking upward into the ore, mucking it into two-ton cars, and sending it out the adit to its destination. As the backs of these substopes rose, square sets were inserted until the back of the stope encountered the floor of the level above and the two previously independent stopes could be locked together. When everything in sight had been stoped out, a large winze was driven down in the most promising spot, and the whole process repeated. The best analogy (and the one the journalists were fond of using) was to compare the process with constructing a Gothic cathedral, spires and all, inside a mountain.

For many years miners walked to work—if descending several hundred feet by ladder could be termed walking. Now and then, when inevitably someone slipped, the usual procedure was to hold an inquest. A particularly grisly feature of these accidents was that the unfortunate did not simply fall straight down to his end but invariably began to ricochet, glancing from side to side of the shaft with ever-increasing velocity until what was left of his body came to rest in the sump. For that reason a shovel and a number of buckets and boxes were always kept

[14] The Chollar-Potosí adit is still maintained as a tourist attraction; the Hale & Norcross portal is visible next door to Archie L. McFarland's assay house below the Fourth Ward School. All the rest of the shafts and adits are gone.

handy at strategic points. At the end of the shift all hands went down to the drain-way level, rode the ore cars out the adit, and then walked uphill to their boardinghouses in Virginia City (Fig. 74).

Milling

In those early years Washoe milling left a great deal to be desired, even after the abandonment of experiments in smelting and attempts to ship out concentrates. Custom quartz mills sprang up like mushrooms, but for long they seem to have employed only a mechanized version of the antiquated *cazo* process. In this process the dressed and upgraded ore was crushed and introduced dry to the battery stamps, which reduced it to dust. The dust was classified (sorted) by air fanning, sufficiently fine particles being collected in a "dust room," whence they were taken in carts to simple pan amalgamators, where water and salt were added. So ineffective was this process that the ratio in ounces of gold to silver recovered was two to three, indicating that by far the greater part of

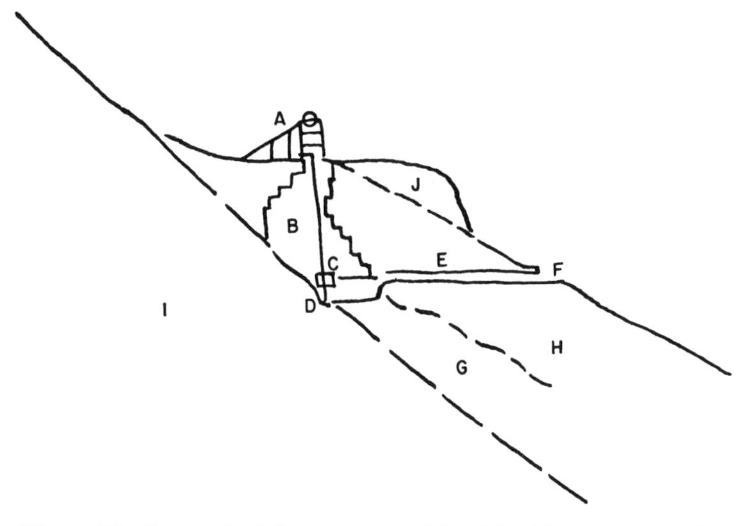

Fig. 74. The Comstock mining system, ca. 1862. *A*, headframe and pumping machinery; *B*, stope (square sets not shown); *C*, pumping reservoir; *D*, sump and pump; *E*, drainageway; *F*, portal; *G*, ore body; *H*, hanging wall (syenite); *I*, footwall (diorite); *J*, waste dump.

the silver sulphurets escaped completely. Needless to say, none of the "processes" improved conditions.[15]

Matters took a turn for the better about 1863, when the quartz mills adopted a version of the patio process. They continued to dry-crush and stamp, but the dust was now introduced to Washoe mullers invented by Israel Knox and Henry Brevoort. Originally these devices were quite small, about the size of a household washing machine, but they rapidly gave way to giants capable of accepting a three-thousand-pound charge of ore. Surrounded by a steam jacket, the bottom of the interior was marked by ridges or corrugations which were rubbed across by a series of shoes, or mullers, mounted upon a circular plate or by arms radiating from a vertical central shaft. One or more handwheels were mounted on top for vertical adjustment of the shoes. A charge of ore, mercury, salt, and copper sulphate was shoveled into the muller, steam was admitted to the jacket, water was added, and the shoes were set to revolving so as to rub more or less heavily over the corrugations. After several hours of abrasion in the muller the shoes were raised clear with the handwheels and thereafter rotated freely as agitators (Fig. 75).

From the muller the pulps were drained by gravity through gate valves into separators. At their crudest the separators were simple, comparatively shallow pans, armed only with slowly revolving paddles. They agitated the pulps deliberately, relying on the principle that amalgam droplets have the fortunate tendency to clump and merge into one another. As soon as a droplet grew large enough that its specific gravity took precedence over the tendency of the agitation to keep it suspended, it sank to the bottom of the pan to join its fellows in a layer of amalgam. In the meantime, a gentle stream of water washed away the top layer of exhausted pulps to keep pace with the entering trickle of fresh pulps from the muller. When investigation indicated the need to halt the paddles and the pulp, the water flow continued, and with some manual assistance the amalgam layer at the bottom was washed clean, removed with the usual spatula, and added to the cleanup for subsequent retorting (Fig. 76).

Both for convenience and as a hindrance to robbery, the mills cast their retort sponge into pigs of one hundred pounds' weight. Melting

[15] Lord, *Comstock Mining and Miners,* 114–29.

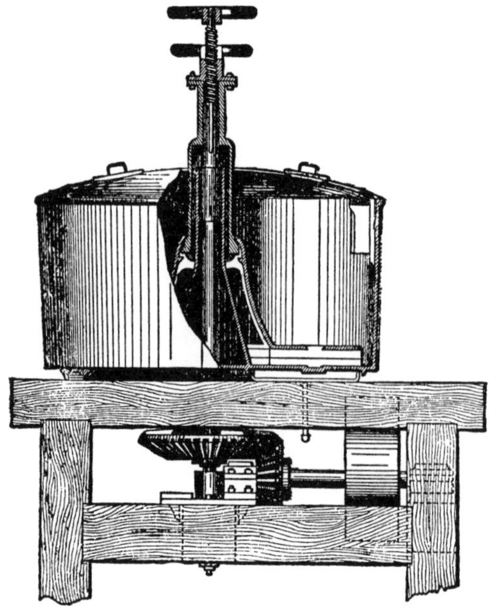

Fig. 75. The Washoe muller. From J. Bernard Mannix, *Mines and Their Story* (1913).

was accomplished in ceramic or graphite crucibles about thirty inches high, with a one-inch wall thickness and with characteristic taper and pouring lip. Specimens are readily found in latter-day mining camps, usually converted into flower planters. The ingots were fire-assayed for fineness by a process much resembling the last part of regular assaying: weighing, fluxing with lead, "parting" of the resultant *doré* bead with nitric acid, filtration, and annealing of the gold particles. But a final step was conversion of the silver nitrate with salt to silver chloride, followed by weighing of the filtrate. The fineness was then stamped on the matte pig along with the mill mark. It can be assumed that in the better-managed mills some effort was made to relate the amount of matte recovered to the day's ore tonnage input as a guide to quality control and as a check on pilferage.

Legal Controversies

By 1861 the Comstock Lode was firmly in business, the headframes

Fig. 76. Comstock silver mill, with mullers and separators. From J. Bernard Mannix, *Mines and Their Story* (1913).

(boxed in as protection from cold and snow) towering above the business houses and residences which adjoined the mine properties, and the mill stacks throwing woodsmoke across the eastern valley (Fig. 77). Shares of the various mines were feverishly traded on the San Francisco board. The wealthier ladies of the city found gambling in mine shares nearly as exhilarating a recreation as they could experience. For those punters of more limited means, there were the penny stocks of the some sixteen thousand valueless gopher-hole claims of the district, whose only attraction lay in the fact that they were undeniably to be found in the Comstock mining district.

With the extent and richness of the Comstock Lode now firmly established, gentlemen of predatory inclinations judged the time ripe to establish dominion over the whole. The first attack relied upon three factors: (1) it was almost beyond dispute that the ore ledges of the lode had been deposited at one time and in one fashion, and hence might be construed to be of one body; (2) under the common law the first claimant (or his assignees) of the major outcrop of the lode was thus

Fig. 77. Waste dumps of the Chollar-Potosi, Savage, and Hale & Norcross mines. From William Wright [Dan deQuille], *The Big Bonanza* (1964).

entitled to all of it; and (3) if any judge or jury was disposed to question this syllogism, the predators had sufficient money and influence to overcome the objections. In December, 1861, the Chollar Mining Company accordingly filed a brief calling for the ejectment of the adjoining Potosí Mining Company from the latter's surface claim. In advancing its plea Chollar, through its attorney, William M. Stewart, argued that the ore which Potosí was hoisting was but a portion of a well-defined ledge whose apex was found and claimed as such by the plaintiff. It goes almost without saying that if the Chollar Mining Company made it stick, it would then proceed to annex its remaining neighbors until, like the walrus and the carpenter, it had devoured every worthwhile property along the entire line of Comstock mineralization.

The alarmed mining companies rallied behind the Potosí group for a fight which came to be referred to as *Jarndyce* v. *Jarndyce*. Potosí's defense took three tacks: (1) it strongly disputed the one-ledge theory with expert testimony; (2) it attacked the priority of lode claims over surface (placer) claims; and (3) it raised a slush fund to purchase witnesses, geologists, experts, and even judges, the established sweetener for the last being twenty-five thousand dollars. For his part, attorney Stewart executed many a neat legal twist and turn against the Potosí group, whom he regarded as claim jumpers and corruptionists. Not at all liking the wholesale auction of their services that was being held almost daily by the territorial judges in the dispute, Stewart rallied the entire Virginia City bar and forced the resignations of Judges James R. North, George Turner, and P. B. Locke. Their replacement by somewhat better men followed swiftly, but although the weight of evidence appears to have rested with the unitary-ledge theory, the Chollar people were unable to gain any definitive decision. Apparently the point had been reached (described by Clausewitz in another connection) in which even total victory was not worth the price that had to be paid to achieve it. The dispute was finally compromised by a merger in which the Chollar and Potosí groups became one, accompanied by a sort of gentlemen's agreement to let well enough alone. Thus by 1864 the first attempt to gain a monopoly of the Comstock Lode had ended in failure.[16]

A second, more insidious, attempt was at once set in train. The ruler of the enormously powerful Bank of California, William C. Ralston, had

[16] *Ibid.*, 131–80.

long been casting covetous eyes upon the Comstock Lode. His bank was always overextended, and possession of the Comstock would not only cure that chronic condition but also allow Ralston to increase indefinitely his assets and his power. In 1864 he opened a branch office in Virginia City. The manager, William Sharon, was instructed to be very generous in his lending policy to all quartz-mill owners who professed need, and even to those who did not. The Bank of California was already planning to build the Virginia & Truckee Railroad to serve the district and controlled the vital mercury market as well. Once a transportation monopoly had been achieved, a milling monopoly would be created either by foreclosure or by dictation of mill policies by the Bank of California. Ralston's methods startlingly resemble those of that other engrosser of mineral wealth of the period, John D. Rockefeller, in that both men saw that control of the product lay in processing and transportation: he who controlled the middle steps would have both producer and market at his mercy. Unlike Rockefeller, however, Ralston insisted on operating alone and in a rule-or-ruin fashion.[17]

The Sutro Tunnel

While Ralston was contemplating the web which he was spinning about the Comstock, his attention was caught by a scheme being put forward by the inspired fanatic Adolph Heinrich Joseph Sutro, who was as dedicated and single-minded an engineer as his great coreligionist, Theodore P. Judah, of the Central Pacific Railroad. As early as 1861, Sutro realized that the water problem of the Comstock must inevitably get worse as the levels descended and that the palliatives of Cornish pumping and drainage adits would presently begin to run into the law of diminishing returns. His proposed solution was both simple and impressive. Noting that the bed of the Carson River was still quite some distance below the lower Comstock workings, he proposed to run a tunnel four miles long from that point under Rose Peak to the arc of the mines, then extend headings laterally north and south beneath the lode. Such a tunnel would not only drain and ventilate all the mines but in addition permit cheap underhand stoping and economic ore removal out the tunnel. The investment plus a handsome profit would

[17] The complete story of Ralston's machinations can be found in George D. Lyman, *Ralston's Ring*.

be realized by levying a modest royalty on the mines so benefited.

Sutro's scheme was technically feasible but met instant and violent opposition from the mineowners. They did not object to Sutro's tunnel as such but did object very much to paying him anything for his services. They reasoned that as soon as they became dependent upon him Sutro would begin charging them all the traffic would bear—which was exactly the sort of thing they themselves would do if they were in his shoes. Rather than let Sutro bail them out collectively, they professed willingness to drown individually. On the other hand, the mineworkers of the Comstock were fascinated by the scheme and wished it well. Sutro even collected a few disciples who were to support him through all the troubles to come with every resource of their minds and purses.

William Ralston was also fascinated by the Sutro Tunnel, seeing it as the final nail in the financial deadfall he was constructing for the Comstock. In his own hands the tunnel would complete his monopoly of milling and transportation, being yet another club to hold over the heads of recalcitrant mining companies who refused to see that submission to the Bank of California was a historical necessity. Therefore, Ralston set in train a scheme of propaganda and obscurantism intended to destroy Sutro's credit and drive him from the region, whereupon Ralston would quietly take over the project, complete it, and use it to forward his own ends. In the decade from 1865 to 1875, Ralston employed every one of the considerable means at his disposal to ruin Sutro, pouring millions in money into those machinations and well-nigh eviscerating the Bank of California in the process. Despite all, the project never quite died. In 1869 congressional approval for the tunnel was obtained, ground was broken for the portal,[18] and three million dollars was raised for the work (Plate 9).

During the course of Ralston's—and Sharon's—stealthy maneuvers, fortune appeared to play into their hands. Mine stocks tended to fluctuate wildly on the San Francisco board, owing not only to covert

[18] The Sutro Tunnel portal still exists and is easily accessible, being but a mile from the highway running northeast from Carson City. Some of the old construction buildings are in fair repair, but the adit itself is badly dilapidated and caved in a few yards from the entrance doors. A small trickle of water emerges from the adit piping, but that and the spoil bank are about all that are left. Apparently no information is available concerning the two hoisting and ventilation shafts which Sutro sank on the flanks of Rose Peak.

manipulation but also to the fact that the Comstock ore, despite its consensual unity of origin, varied considerably in tenor: one mine might be hoisting bonanza ore while its immediate neighbor, stoping at the same level, would be temporarily in *borrasca*. Each rumor of fire, flood, or *borrasca* depressed stock prices; each announcement of discovery of a high-grade block of ore inflated the prices of stocks in all mines. At each assault of the bears, Ralston quietly bought. When the market was bullish, he sold enough stock to finance his next foray. He gained control of the great flume system which brought in the square-set timbers, completed the Virginia & Truckee Railroad so as to drive the freighters out of business, and, master of Machiavellian tactics, let Sharon do the dirty work of foreclosure and annexation. His procedures did not enhance Sharon's popularity, and when Ralston proposed to purchase from the notoriously venal Nevada legislature William M. Stewart's seat in the United States Senate for Sharon, the *Territorial Enterprise* fired a broadside of excoriation:

> Your unexpected return, Mr. Sharon, has afforded no opportunity for public preparation, and you will consequently accept these simple remarks as an unworthy but earnest expression of the sentiments of a people who feel that they would be lacking in duty and self-respect if they failed upon such an occasion to make a deserved recognition of your acts and character. You are probably aware that you have returned to a community where you are feared, hated and despised. . . . Your career in Nevada for the past nine years has been one of merciless rapacity. You fostered yourself upon the vitals of the State like a hyena, and woe to him who disputed with you a single coveted morsel of your prey. . . . You cast honor, honesty, and the commonest civilities aside. You broke faith with men whenever you could subserve your purpose by so doing.[19]

Nevertheless, Sharon got his seat. And, as an afterthought, Ralston also purchased the *Territorial Enterprise*.

During the course of those ten years the mines went deeper and in so doing were for the fourth time forced to revise their technology. It will be recalled that they originally worked by open-cast methods, then drifted by means of inclines. When the dip became vertical, inclines were abandoned in favor of shafts and connecting hillside adits. About 1866 the ledges were found to change "hade" again, dipping eastward

[19] Lyman, *Ralston's Ring*, 213.

and following the wall rock of Mount Davidson diorite. The mines therefore moved a thousand feet east and down the surface slope to sink entirely new hoist shafts vertically through the country rock and the ore bodies until they hit the footwall. This work was admittedly unprofitable for the moment, but the ground was now reasonably free of water, and, although as "heavy" as ever, could be timbered well enough. When the footwall was reached, an incline was begun, following the lower horizons of the ore body. It must have been maddening work, since it was almost impossible to dewater the headings of the inclines—Cornish pumps were hard to make turn a second corner, the levels were too low to benefit from the customary hillside drainways, and the Sutro Tunnel had not yet broken ground. The solution was to leapfrog by running drain ways to neighboring mines, so that while one was advancing its heading the other would pump for both. Such co-operative agreements had earlier been worked out for ventilation, fire control, and safety.

Once the incline was finished, crosscuts went out from the new shaft into the ore body, square sets were installed, and overhand stoping was begun. The ore, as before, went down chutes and up winzes to the incline, where the chute tappers ran it into an oversized five-ton tramcar called a giraffe, which had gained its name from the fact that its uphill wheels were considerably smaller than its downhill wheels, permitting its bed to set level despite the inclination of the passage. The giraffe was winched about by cable from the surface enginehouse, and every observer was careful to point out that this cable ran *under* the sheaves which were installed at the bottom of the vertical portion of the shaft.[20] The giraffe then deposited its load in the cars of the hoist cages, which were now double- or triple-decked (Fig. 78).

Despite the new approach, by 1869–70 the Comstock appeared to be going again into *borrasca* as the lower ore horizons thinned out slowly but consistently. Between exhaustion of the major ore bodies and the rising costs of battling the water, shares in the better producers were dropping sharply. The speculative stocks in the gopher-hole mines, whose sole value lay in their physical proximity to the big mines, slumped until control of them could be bought for a pittance. Ralston and Sharon were not interested in them. But another man was: John W. Mackay,

[20] Wright, *The Big Bonanza*, 120–22.

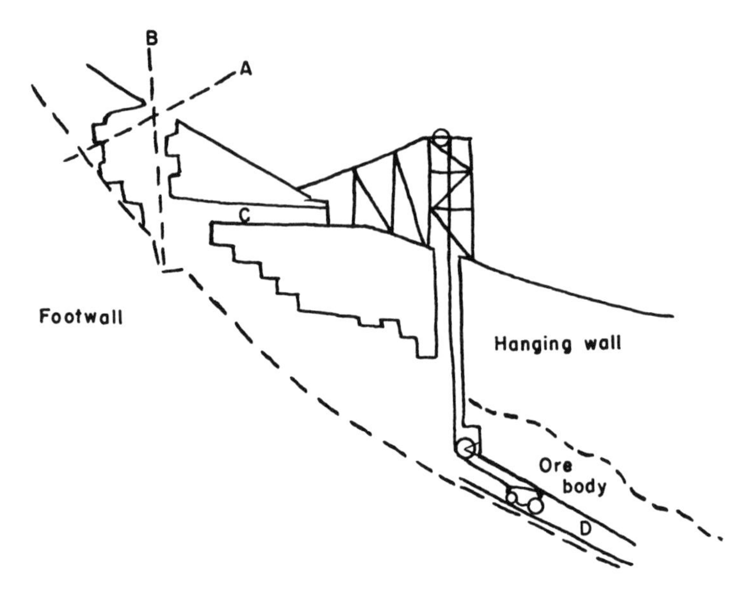

Fig. 78. The Comstock mining system, 1866 (not to scale). *A*, axis of original incline; *B*, axis of abandoned shaft; *C*, old hillside drainage adit; *D*, new haulage incline and giraffe car.

who scented opportunity. Mackay was a hard-driving mine superintendent who had the reputation of being able to smell sulphurets through ten yards of porphyry and who seemed to have proved it by his work in the Kentuck Mine. His attention became directed to the fact that between the Best & Belcher and the California mines, both of which had been good producers in their day, a short segment of the Comstock arc lay totally neglected. To be sure, some trifling exploratory work had been done in the gap, but to that time none of it had gone below the six-hundred-foot level. Indeed, the Virginia & Truckee Railroad now had its yard and depot on the surface of the site.

Mackay reasoned that a hidden ore body of a size at least equal to its sometime very profitable neighbors lay virtually under the Virginia & Truckee yards, but at a depth which had allowed it to escape previous investigation and in a lobelike form which isolated it from the deep stopes of its neighbors. Logic (if geology ever conforms to a priori logic) indicated that his theory had to be correct, since there was silver ore continuously along the arc except for that one spot.

At that time the stock of the wildcat claims within the gap could be bought very cheaply. Mackay talked so convincingly that he attracted partners: James G. Fair, who had made the low-grade Hale and Norcross Mine pay well, contrary to all expectations; and two San Francisco high rollers, William S. O'Brien and James C. Flood. They pooled their assets, gained control of the surface claims, and sought permission from William Sharon to begin drifting north from the 1,167-foot level of his Gould and Curry workings. Remarking that he would be glad to "help those Irishmen lose some of their money," Sharon assented. Their superintendent, Captain Sam Curtis, began the heading, drifted north under the Best & Belcher, then turned northwest toward the wall rock of Mount Davidson. On September 12, 1872, the heading intercepted a 7-foot fissure crammed with broken syenite, clay, and low-assay quartz. Assuming that it was a shoot from the major deposit which he sought, Curtis began to follow its northeasterly strike.

As he had predicted, Curtis found the fissure widening and increasing in assay value, enabling the partners to pay off their debts, buy more outstanding stock, and sink an independent shaft of their own. As soon as the new shaft holed through into Curtis' exploratory drift, the Big Four installed machinery and commenced hoisting twenty-three-dollar ore from the forty-eight-foot ledge which had now been revealed. In February or March, 1873, the top of the long-sought major deposit was revealed: a lode twenty feet high, forty feet wide, and going from $50 to $400 in silver per ton.[21] Peculiarly enough, it occupied a fissure in the hanging wall of the Comstock structure; every other major deposit

[21] Smith disposed of the following persistent legends: (1) that James Fair was in direct charge of the exploratory work and was the man who traced and found the lode; (2) that the thin seam of sulphurets, often no more than a film of clay, which Fair "followed like a blood-hound," was the only lead to the main ore body; and (3) that the exploratory work was conducted in high secrecy until the day when Deidesheimer and Wright were invited down to view the lode and publish the news. Smith showed that Curtis was in charge of the exploratory work, that the lead to the main ore body was quite obvious, and that the newspapers were given the full facts as exploration progressed. The first two legends appear to have been put forward by Fair himself, and the last was based on a Comstock custom known to have been employed in other mines, but not in this particular development (*The History of the Comstock Lode, 1850–1920*, 148–54).

had lain along the footwall. Both Deidesheimer and William Wright were invited down to take a look. They proclaimed the glad tidings far and wide, and it was very optimistically predicted that no less than $300 million in ore was in sight. The good news was not good for Ralston; he had gone short in Ophir shares and was caught in the general bull market which followed the announcement. Suspicion of what was afoot caused a run on the Bank of California, and when Ralston's accounts were examined, well over a million dollars remained unaccounted for. That evening Ralston drowned under murky circumstances.

The apparent revival of the Comstock caused Adolph Sutro to redouble his efforts to complete his tunnel. He introduced air drills in 1874 and on July 8, 1875, fired the final blast at the end of 20,100 feet of remarkably difficult heading that brought him out under the Savage Mine, two claims south of the Consolidated Virginia. The war for the Comstock now narrowed down to a struggle between Sutro and the intransigent mine companies. Water proved to be Sutro's ally, and one after another the owners capitulated in the face of the rising floods. Sutro salved their anguish by making concessions in his royalty price of two dollars a ton but, seeing the handwriting on the wall for the entire district, quietly sold his equity in the tunnel and went back to San Francisco to engage in good works. Most of the mines were played out, virtually all were now below the level of his tunnel, and it was apparent that it had been finished too late to be of economically important service.

The Con Virginia Mill

Despite the omens, the Consolidated Virginia had enough proved ore to justify installation of a quartz mill that was the pride and wonder of Washoe. Since it was the last and greatest of the empirical silver mills, an outline of its operation is in order (Fig. 79). Though not the largest of the district mills (that of the California Mine was the champion), the Con Virginia mill[22] boasted sixty stamps with a daily capacity of 250 tons of ore. Erected, as a proper mill should be (discounting

[22] Though it was so called, it was built and operated by a joint subsidiary company, the California Company, of the Con Virginia and the California mines (*ibid.*, 159–60).

Fig. 79. Gold Hill (top) and the Consolidated Virginia headworks (below). From William Wright [Dan deQuille], *The Big Bonanza* (1964).

the risk of fire), only two hundred feet from the hoist shaft of its mine, it stood on a terraced grade; the ore went in at the high end, and thus virtually all of the process conveying was done by gravity. Mules pulled trains of ten ore cars at a time from the headframe platform to the ore bins—apparently these mules were less intelligent than the Bisbee, Arizona, breed—where the ore was dumped down a chute to a three-inch grizzly or grating. Anything that passed the grizzly went directly to the feed bins, while anything larger was shunted aside to a substantial jaw crusher.

The dressed ore was fed automatically by Tulloch self-feeders (Fig. 80) to the stamps on a demand basis. A glutted stamp disengaged itself from its lifting cam and simultaneously shut down the feeder mechanism until the ore beneath its shoe was sufficiently diminished to re-engage both. The mill introduced two sensible innovations in the stamping process: it wet-stamped the ore, thus eliminating dust, and the ore itself was fed parallel to the line of stamps so that it made a pass under several before trickling out the screen at the end of the battery. These pulps flowed down to large settling pans, where surplus water was eliminated. When the pans were full, they were emptied by hand, and the ore was stacked on a platform, whence it was shoveled

Fig. 80. Tulloch automatic ore feeder. From Joseph C. F. Johnson, *Getting Gold* (1917).

into the Washoe mullers, of which the mill boasted thirty-two, each holding a three-thousand-pound batch.

Another innovation in the Con Virginia mill was its mercury system, consisting of a central reservoir tank, iron piping, and a feed pump whose valves were solid India-rubber balls. Mercury was run into the mullers merely by opening a valve, an arrangement which eliminated the customary spectacle of mill hands staggering about with seventy-six-pound flasks clutched in their arms and either spilling several pints of valuable mercury over the floor or, as occasionally happened, letting the entire flask fall bodily into the muller, whence somebody had to fish it out before the muller could be put back "on line." The Con Virginia people were enormously proud of their mercury system and apparently lost no opportunity to show it off to visitors and to brag about it.

Mulling was an intermittent process that required five hours; the management inexplicably ran the machines half that time before adding the mercury and patio reagents. It would have been far more sensible to add them to the ore right in the stamp batteries, but somehow the millmen of the period never caught on to this practice, which was standard with the supposedly backward Spaniards. When judged finished, the pulps were run down into the separators, which were highly sophisticated for the day. Instead of being cleaned by hand, each separator automatically drained the settled amalgam through a pipe in the bottom leading to strainers.[23] The strainers were locked iron cases enclosing conical canvas bags. Superfluous mercury dripped out through the fabric through force of gravity and was run back to the central reservoir for recycling. When a strainer bag was full, the case was unlocked by a supervisor, the loaded bag was put in a rail car to be pushed to the retorting house, and an empty bag was slipped into the holder.

Meanwhile, back at the separator line, the pulps and water were drained down to one of the four great agitators in the lowest level of the mill building. The agitators were huge tubs equipped with revolving rakes; actually, they were another type of separator designed to reclaim amalgam which had escaped the first run and in addition

[23] The quantity of mercury introduced to the mullers must have been proportionately greater than usual, in order to afford a rich mixture that would flow more readily than the usual mill practice found desirable. Crisper and leaner amalgam would not have drained as well.

served as settling tanks to drain yet more superfluous mill water off the top of the contents and reclaim it. From time to time the agitators were cleaned out (probably by careful shoveling and hosing), the exhausted pulps being sent to their final repose on the tailing heap below the mill and the heavy bottom layer of amalgam and detritus being recycled through the separator line.

The strainer bags were assembled at the hydraulic press, where their contents (resembling gray mud) were dumped into the mortarlike device and subjected to direct water pressure of 150 pounds per square inch. This increased the metal-mercury ratio from the "natural" one to seven to a more economical one to four, with the mercury squeezings going, as before, back to the main quicksilver reservoir. The slugs of amalgam were put back in the same car and delivered to the retort furnaces, where the mercury was distilled off (and delivered automatically from the condenser to the reservoir), while the sponge matte was melted and cast into bars in the adjacent melting room. These bars were assayed on the premises before delivery by railroad to the United States Mint at Carson City. Although the Con Virginia, following good precedent, was not fond of releasing production figures, it let slip that in the heyday of production it was retorting from 2.0 to 2.5 tons of amalgam a day. Some tentative arithmetic based on these figures fixes daily matte production at a generous 1,250 pounds, of which 10 per cent was gold, with a total nominal value at current prices of about $40,000. Assuming that the mill was working at its rated capacity of 250 tons of ore, one concludes that the mine-run ore of the Con Virginia went a bit less than $200 a ton, a figure which should dispel the rather common supposition that the ore was virtually solid silver.[24]

It is impossible to estimate milling losses, but they were known even at the time to be considerable still, despite the fact that the Consolidated Virginia mill was using the patio process in the most efficient plant the technology of the day was capable of constructing. It was accepted as sober fact that the Carson Sink was full of sulphurets which the early *cazo* process millers had missed completely and let go as tailing. There were losses in other directions as well. As an experiment, one mill superintendent set a copper bowl and some copper riffle bars, all coated

[24] Basic figures on ore and amalgam production are from Wright, *The Big Bonanza*, 254–67.

with mercury, under the flume from which his mill water emerged. The water appeared to be crystal-clear, but at the end of three months, a cleanup of the bowl and the bars netted over one hundred dollars.[25] Strangely enough, there seems to be no evidence that this discovery was taken advantage of on a systematic basis.

The Decline of the Comstock

The story of the Comstock Lode trails off abruptly after the discovery of the Con Virginia and the completion of the Sutro Tunnel. Mine after mine was first weakened structurally and financially by fire or flood, then hit *borrasca* in its lower levels, and finally exhausted its financial reserves in attempting to find upper-level ore shoots which had been overlooked or disregarded during the bonanza times. Diamond-drill probing revealed nothing that would pay the geometrically increasing costs of delving deeper and pumping higher. In that period of decline the mines were thoroughly cleaned out. Labor costs were low, and nothing was ignored so that, despite persistent rumor and stock puffing, there is really no silver left in the Comstock worth going after.[26] From time to time developments which may have been serious or perhaps intended only to enable shareholders to unload gave rise to temporary flurries of interest, but by about 1900 the original Comstock Lode was dead.

During the next twenty years cyanidation created a new area of activity on which there are few available production figures but which was undoubtedly quite profitable. The ability of cyanide to extract the last particle of gold from tailing or low-grade ore inspired the erection of dozens of cyanide mills in the Virginia City–Gold Hill area, a good many of which are still visible, whereas all the old-time quartz mills have vanished completely. The low-tenor gold shoots of Gold Canyon, which had given rise to the original Johntown placers, were systematically mined out. The tailing heaps of the original mills were fastened upon, run through the vats, and pumped away. Therefore, although the original waste dumps are still visible in place, the tailing heaps which once were found below them are gone. An example of this activity is to be seen below the well-marked site of the Rhode Island Mill.

[25] *Ibid.*, 96.

[26] Interview with Archie L. McFarland, assayer of Virginia City.

A cyanide scavenger moved in, erected a dragline with a clamshell bucket to seize the Rhode Island tailing, cleaned it out, and dumped it back in the Gold Canyon stream bed, where it is now visible as a dead-white flat of detritus. Vegetation is only now beginning to creep over this flat, a generation after the last bucket was dumped.[27]

Little else remains. C Street of old Virginia City is still fairly intact, lined with shops catering to tourists. A few hundred feet down the steep slope of Mount Davidson are the yellow waste dumps of syenite, but the shafts have been filled in, the adits have collapsed, and few of the sites are even marked. A clean sweep has been made of the original mills and machinery, even to the granite foundation blocks; only the sheet-iron-covered structures of the later cyanide epoch are visible. It is difficult to envisage that from this surprisingly restricted valley came over 375 million dollars in silver and gold[28] to affect the destinies of men, political parties, and nations. It is even more difficult to realize that beneath the brushy slopes still lie the lightless stopes, littered with rotting timbers and rusty iron, their eternal silence broken only by the drip of scalding water.

[27] Whitman Symms, "Decline and Revival of Comstock Mining," *Mining and Scientific Press*, Vol. XCVII (October 10, 24, 1908), 496–500, 570–76.

[28] Smith, *The History of the Comstock Lode, 1850–1920*, 293, 297. This figure is the estimated production from 1850 to 1920. For the flush times of 1850–82, Smith gives total production as 300 million dollars. Dividends amounted to 125.3 million dollars, but assessments totaled 75 million dollars. Thus, over-all profits were but 50.3 million dollars, and the vast bulk of these profits were derived from only eight or ten mines. Some of the most renowned—the Ophir and the Mexican, for example—showed a net loss over the long run, while others, such as the Utah, produced no profitable ore at all despite costly development.

CHAPTER X

Gold, Silver, and Cyanide

EAST OF THE CENTRAL ROCKIES and separated from them by some two hundred miles are the Black Hills, so called because of their dense cover of evergreen forestation, which appears green-black from a distance. During the Tertiary period unknown causes upraised the area tectonically in a pistonlike manner, giving the region the appearance of an enormous dome. The exposed central core is Harney Peak granite, a light, almost creamy rock which was formed as a massive Precambrian batholithic stock. Around this core the formations are concentric, granite giving way to schist, and the schist to a limestone which is the country rock. Deep washes radiate from the center, making the region difficult to traverse. Since flat prairie surrounds the Black Hills on all sides, early travelers had little incentive to poke about in them. The Sioux Indians also discouraged such activity, so that, although the Black Hills were on occasion visited by white men from the days of La Vérendrye in 1743, none felt any compelling reason to linger in the vicinity or to explore its streams and gulches.

The Black Hills Gold Rush

Beginning in the 1830's there were vague but persistent rumors of gold to be found in the Black Hills. The prospectors of the big rush of 1874 kept turning up physical evidence which proved (at least to their satisfaction) that people interested in gold had been there quite some time before. This evidence ranged from the Thoen Stone, whose circumstantial tale of gold and bad Indians in 1833 may be credited (by those who wish to do so) to reports of fair-sized trees growing from the bottoms of long-abandoned prospect pits. The hills are certainly known to have been prospected, and some gold was recovered in 1863 by the G. T. Lee party. But many factors, including the Civil War, the unfriendly attitude of the Indians, the existence of more attractive mineral districts elsewhere, and difficulties of transportation discouraged more intensive displays of curiosity. The end of the war and the completion of the transcontinental Union Pacific Railroad brought the region within reasonable reach of good transportation and marked the end of the Black Hills' isolation.

Another factor in the slow start of activity there lay in the fact that prospectors are oftentimes reluctant to enter new terrain, preferring to confine their activities to regions which are known to be mineralized and whose signs and indications they consider themselves to be proficient in reading. A new district implies learning new eccentricities from the ground up, with no certainty of any return for the effort involved. By the same token, sale of a claim in unproved territory is attended with difficulties, mining syndicates desiring established bases and communications, as well as some definite ideas (gained at someone else's expense) of probable extent and tenor of ore. In short, nobody greatly cares to be the first in a new mining district, although everybody yearns to be the second, as witnessed by the rushes which come hard on the heels of a major strike, or even on the rumor of a strike.

The propensity for rushing to a new district was regarded humorously by the prospectors themselves, who were fond of a campfire tale involving an old desert rat who died and applied for admission to heaven. Saint Peter (goes the tale) was inclined to turn him away on the grounds that heaven already had entirely too many prospectors.

"That's all right," said the old-timer. "Leave it to me, and I'll thin them out."

Being admitted provisionally, he at once began to talk about a major strike that had been made in the infernal regions, with the result that in a few days most of his colleagues had packed up and left. Saint Peter noticed, however, that the old prospector did not seem to be elated—that, in fact, he seemed morose and thoughtful.

Soon the prospector came up to the Keeper of the Keys and burst out, "Peter, I guess you'll have to let me out of here. You see, on thinking it over, there may be something in that report of rich diggings in hell."[1]

After Appomattox several associations were formed in Dakota Territory to subject the Black Hills to intensive search. These ambitions ran head on into the new Quaker Peace Policy, which had been adopted by a national administration interested only in peace, retrenchment,

[1] Hughes, *Pioneer Years in the Black Hills*, 136.

and profit taking. The active sponsor of this policy, the Society of Friends, had long held the theory that if the American Indian was granted inviolable reservations and was sufficiently well armed to make peaceful hunting relatively easy, then "Lo, the poor Indian" would no longer study war, neither would provocation exist for war to be made against him. Since the Black Hills were well within the Sioux treaty lands and were held by the Sioux to be of great medicinal value, the Quaker-dominated Indian Bureau brought great pressure upon the Grant administration to discourage interlopers of every sort.

The Quaker Peace Policy seemed so ethical, so logical, so economical to politician and true believer alike that some of the objections to its feasibility were brushed aside as irrelevant. Yet one of those objections lay in the enforcement of the key word "inviolable." There was no federal constabulary in unorganized territories save the United States Army and Congress had always refused point-blank to grant the army any police power or to establish courts before which offenders could be tried for violations of Indian policy. The result was in direct contrast to what occurred in Canada, where a highly efficient constabulary, the Royal Canadian Northwest Mounted Police, and no-nonsense crown courts elicited a great measure of respect for law among whites and Indians alike. It was questionable whether the United States Army could arrest civilians in the absence of proclamation of martial law by the president. But even if it could, it could not try them by court-martial, and the question was moot whether it could cite them before territorial or any other federal courts. The Supreme Court decisions in *Ex parte Vallindigham, In re Milligan*, and so forth, just about limited the army's powers to admonishment or bluff. There was no doubt that a military officer who overstepped a vague and uncertain boundary could be sued for damages in his private capacity and that the court of first instance would probably view with cynicism his plea that he was acting in obedience to a military order.

Military power over the reservation Indian was almost equally limited, save when the War Department could be persuaded to declare a given band hostile and thus subject to military coercion or arrest. The Indian Bureau was singularly reluctant to acquiesce to such declarations, since the Friends who administered the bureau were persuaded that the army was only seeking an excuse to kill some of its wards. Now and then a strong-minded and politically potent officer like Sheridan could cut through the red tape with impunity, but the average military officer had little political clout and did not relish the prospect of being forced to pay a court judgment or even cashiered for his pains. In short, the inviolable-reservation policy was unenforceable since Congress did not really wish it to be effective and would not grant the means to make it effective.[2]

An accessory to this situation was the Indian Bureau's insistence on presenting ultramodern repeating rifles to the delighted Indians. Indignant army officers contended that the Indian was a warrior in manners, morals, and politics and that if he could not find war enough to content him at home he would travel abroad to seek it. The bureau waved aside this objection as distortion of true facts well known to the bureau: the announced peaceful intentions of the chiefs with which it had parleyed and the Friends' dogma that no reasonable human being would willingly make war, particularly when counseled by those having possession of the Inner Light.

This policy was cast into involuntary bankruptcy by political pressure from the Dakota Territorial Assembly, which early in 1873 demanded a military-scientific survey of the Black Hills for the benefit of gold seekers. This demand was coupled with the thin rationalization that the hills served only as a refuge for bad Indians, who would be denied its benefits by white settlement. The Seventh Regiment of Cavalry, commanded by Lieutenant Colonel George A. Custer, was detailed to escort surveyors and investigators in 1873 and 1874. It was evidently hoped that the gold findings would be scanty, the impending boom forestalled, and everyone, Sioux included, made happy without difficulty or cost. As was the case with nearly every program the Grant administration launched, the scheme backfired. Some interesting colors were found by the old prospector Horatio Nelson Ross on August 1, 1874, and Cus-

[2] *Ex parte*, this legal tangle casts a somewhat different light upon such strictures as those of the late Mari Sandoz, who once generalized that the Indian "could always trust the Red-Coat [the Canadian Mounted Police] but never the Blue-Coat [the United States Army]." In point of fact, the Canadian constabulary had the power to enforce the law and reservation policy, but the United States Army was consistently made the scapegoat for situations which it had not created and was denied the power to cure.

ter's scout, "Lonesome Charlie" Reynolds, greatly magnified the discovery in the newspapers. Over the threats of the Sioux and the anguished protests of the Indian Bureau, a rush began, spurred by the national bank panic of 1873. For a time the army tried to stem the tide, but the administration finally threw in its hand, at once breaking faith with the Sioux and announcing that gold seekers might enter the hills, but at their own risk. This feckless evasion was paid out in the Sioux War of 1876–77, and though it is not recorded that any Indian commissioners were shot by the rifles they had given the Sioux, other individuals may have suffered some inconvenience.

In the Black Hills themselves there was much unprofitable scrabbling about before the placer diggers and miners became aware of the nature of the mineral deposits and the appropriate techniques for working them. Placering, which was centered about the settlement of Deadwood, revealed that the pay streak was seldom or never at bedrock but was ten to twenty feet above it, on a sort of false bottom of compact, clayey gravel, very hard to excavate and wash. The best signpost of the pay streak was garnet, whose pink sparkle even the many greenhorns in the diggings could detect relatively easily. Coyoting was employed extensively, but the holes were so wet that many had to be continually hand-pumped at great effort.[3] Often enough the pay streak was so consolidated by pressure and cemented iron salts that it had to be drilled and shot. This iron and accompanying manganese so coated the gold that amalgamation was greatly hindered. The solution achieved was to treat this "cement dirt" like ore, running it through a stamp mill to crack the surface film and expose the gold to the mercury.[4]

The Homestake Mine

Although many mineralized quartz veins were found, it finally dawned upon the geologists that they were comparatively recent in origin and had little to do with the bulk of the placer gold. In fact, the presence of clay, iron, and manganese indicated that the placers had

[3] A mechanical variation was the wheel and China pump, a chain-pump wheel extending into the sump of the prospect pit, run by a flume-impelled overshot wheel. Since putting the two on a common axle would have entailed building them awkwardly large, the supposition is that the two were connected by belting or an equivalent arrangement (Hughes, *Pioneer Years in the Black Hills*, 141–42).

[4] Watson Parker, *Gold in the Black Hills*, 54–56; Lakes, *Prospecting*, 238–40.

been eroded from the dissemination structure of the central granite core of the hills, then concentrated in the false bottoms of the radial washes. Since the general surface trend of the hills was southeastward, it followed that the chief belt of the diggings was on the southeastern side. The quartz intrusions, on the other hand, were independent in time and deposition of the Harney Peak batholith, although that did not discourage prospectors from fanning out to look for them.

Thus it happened that, in the spring of 1876, a few miles southwest of Deadwood, a party of four prospectors was working the Golden Terra, one of the quartz intrusives. The party consisted of the brothers Fred and Moses Manuel, Alex Eng, and Henry Harney. Reported Moses Manuel:

> Toward spring, in the latter part of March or April, four of us found some rich quartz. We looked for the lode, but the snow was deep and we could not find it. When the snow began to melt I wanted to go and hunt it up again, but my three partners wouldn't look for it, as they did not think it was worth anything. I kept looking every day for nearly a week, and finally the snow got melted on the hill and the water ran down a draw which crossed the lead, and I saw some quartz in the bottom and the water running over it. I took a pick and tried to get some out, and found it very solid, but I got some out and took it to camp and pounded it up and panned it and found it very rich. Next day Hank Harney consented to come and locate what we called the Homestake Mine, the 9th of April, 1876.[5]

The partners began development work, selling the Golden Terra to finance their discovery shaft and a small ten-stamp mill. Development revealed that the outcrop was a steeply dipping ledge of quartz conglomerate which was thought to have originally been the surface of a very ancient ocean beach or low island. It was quite extensive, and seemed to be shot through with "fossilized" placer gold presumably eroded from the central batholithic formation when it was exposed in some previous epoch.[6] Letting it be known that the Homestake and the related Old Abe were for sale, the partners awaited buyers.

[5] Rickard, *A History of American Mining*, 213–14.

[6] Lakes, *Prospecting*, 234. Later research invalidated this supposition. The Homestake resembles the Witwatersrand in genesis, being composed of a Precambrian quartz conglomerate much tilted and metamorphosed. The values came from below by igneous intrusion in both cases, however, as shown by the fact that gold surrounds particles of unaltered pyrite. Had this pyrite been exposed to the at-

A big Comstock Lode operator, George Hearst, and his partner, James Ben Ali Haggin, grew interested in the possibilities of the Homestake claim. After obtaining a report on it, Hearst decided to gamble that the ledge ran deeply and was of comparatively uniform gold content all the way down. The very thinness of the assay value was a hopeful sign. He also liked the formations roundabout. The presence of schist hinted that the lode had been submerged and preserved by sediments in the almost erosion-proof environment of ocean bottom. The final upraising of the Black Hills had steeply tilted the formation and metamorphosed the sediments into schist which was sufficiently waterproof to have kept ground water from playing its usual tricks. Hearst and Haggin bought the Homestake for a trifle over $100,000. The measure of their good judgment can be estimated in the fact that from 1878 to 1962 the Homestake Mine produced $715,000,000 in gold bullion and remains today the last and sole producing frontier-period gold mine of any real consequence within the continental United States.[7] Of course, the fact that the Homestake was a dry mine and that its gold was virtually free-milling helped profits considerably, too.

Milling machinery in the Black Hills was virtually identical to that used in the Comstock Lode, although neither the placer dust nor the free-milling Homestake gold required much treatment beyond crushing, stamping, and amalgamation. The Homestake mill, however, had an improved stamp battery, or mortar, which was lined with silver-plated amalgamation plates, and the sensible procedure was inaugurated of adding amalgamation mercury to the ore right in the battery where it was being wet-stamped, a process called inside amalgamation.[8] It is reported that the Homestake mill was content with these stamps right up to the 1930's. It has also been stated, though without sufficient authority to be taken too seriously, that mine-run Homestake ore goes but one-half ounce of gold to the ton.

Chlorination

Unlike the placer lodes, the Black Hills Tertiary quartz lodes were

mosphere, as would be the case in a true "fossil" placer, it would have decomposed. Cf. F. H. Hatch, "The Conglomerates of the Witwatersrand," in *Types of Ore Deposits* (ed. by H. Foster Bain), 218–19; Rickard, *A History of American Mining*, 217–18.

[7] Parker, *Gold in the Black Hills*, 196–97.

[8] Rickard, *A History of American Mining*, 218–19.

pyritic and refractory. A new milling departure was attempted in the introduction of the chlorination process in one of its infinite variations. Actually, chlorination was medieval in its roots, going back to the alchemists' discovery that aqua regia (a mixture of concentrated hydrochloric and nitric acids) was the only chemical in their laboratories which would dissolve metallic gold. It was also noted along the way that common salt (sodium chloride) was a good thing to use in milling, although no one knew just why, or at what point to use it. There matters rested for a long time, with salt being used everywhere but aqua regia not at all, since it would dissolve gold, to be sure, but would also attack everything else from concrete tanks to millmen's pants. Empirical chemistry was sufficiently far along by the 1870's to ascertain that the critical element was the chlorine; gold would react with it to form the soluble chloride. For that matter, the old Freiburg process had been a form of silver chlorination. Now chemical treatment was revived, with the chlorine being supplied from bleaching powder or as a gas obtained by heating common salt with sulphuric acid. The gold chloride thus achieved could then be drained off as a solution and reduced with virtually any common base metal.

For a time, then, chlorination was touted as the magic key to recovering thin values or milling losses, and chlorination mills blossomed briefly in the Black Hills and throughout the rest of the West. They were almost as soon abandoned, since experience demonstrated that chlorination was not nearly as effective as theory indicated. The gaseous process required large-capacity, pressure-tight reaction tanks, costly to build and difficult to clean out when it came time to remove the exhausted tailing (Fig. 81). Bleaching powder was expensive in the quantities required, and the gas was so dangerous as to be profitably employed as the first of all chemical-warfare agents a generation later. Finally, most metals will react with chlorine before gold will, meaning that the pregnant liquor drained off was heavily contaminated with base-metal chlorides and required expensive additional treatment. After one use, the chlorine reagent was lost for good. In short, chlorination as then attempted was an ideal example of everything that sound industrial chemistry should not be. The practical millmen soon woke up to this and relegated chlorination to the place whence it came: use in preliminary silver ore dressing, wherein salt or some other chlorine com-

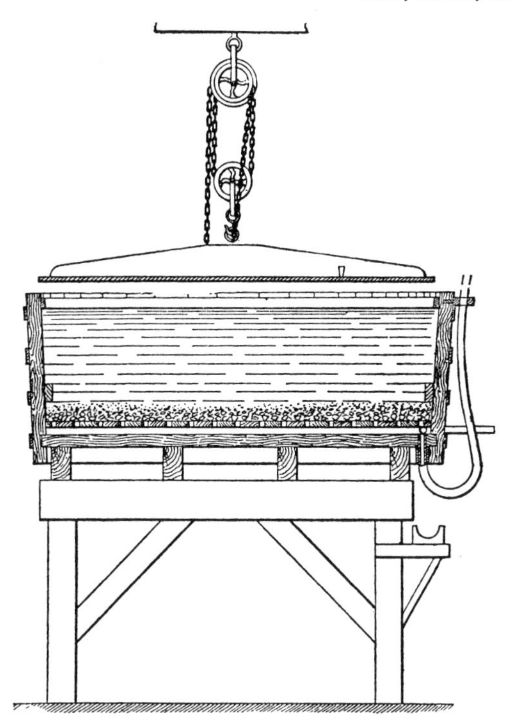

Fig. 81. Chlorination pressure leaching vat. From Thomas Egleston, *The Metallurgy of Silver, Gold, and Mercury in the United States* (1887–90).

pound was roasted with the ore to convert some of the sulphurets to the more amenable chlorides.

Tombstone

While the Black Hills boomed, prospectors continued quietly to rake over the southwestern deserts and mountains. One such was the prospector Ed Schieffelin, who drifted from California into Arizona Territory in mid-January, 1877, with an idea of prospecting in the Grand Canyon. One glance at the undisturbed sedimentary strata above and the barren schists below was enough for Ed; he wandered southeast through the central mountain belt, finding nothing. By spring, 1877, he was so far down on his luck that he went to work as a guard for the crew doing assessment work at the Brunckow (which Ed always called the "broncho") Mine in the far southeastern corner of the territory. The region included the Dragoon Mountains, the hideout of the Apache chieftain Cochise. In addition, the area afforded a good war road for the White Mountain Apaches, who regularly went south through it on their way to raid the ranchos and jacales of Old Mexico for horses and women; the military had sited Fort Huachuca there specifically to discourage such raids. For very good reason, therefore, Schieffelin's tools of trade were field glasses and a rifle, the one to observe war bands working through the desert hills and the other to prevent them from speculating at the expense of the Brunckow Mine crew.

The work was light but boring and with time hanging heavily on his hands, Schieffelin began examining the country. He presently focused his glasses and his attention upon a weathered, bleached, and much-distorted gray-granite plateau northeast of his lookout station. Inasmuch as the higher edges of the plateau were comparatively sheer and exposed, Schieffelin was able to observe faults, synclines, and other evidence that the plateau had been heaved and much mistreated subsequent to its first uplift. Since he was virtually sitting upon a demonstration that the region was mineralized, the plateau seemed to him an excellent place to prospect. Presently Schieffelin threw up his job and went to Fort Huachuca to arrange for a grubstake and to lay in some supplies. The soldiers there warned him that he was likely to find nothing but his grave in a region so heavily infested with Apaches, but the Gold Bug had its fangs buried firmly in Ed. He went forth in company with the assayer Richard Gird and his own brother Al, to work down the San Pedro River, which skirted the plateau he wished to investigate.[9]

Clambering about the rim, Schieffelin found blossom and outcroppings which Gird's assays showed to have silver values of high tenor. Those areas he staked as the Graveyard and Tombstone lode claims. Working around to a natural amphitheater on the southeast, he found a heavy stain of iron spreading downward fanwise—a most excellent and promising sign. What he found at its apex was not, however, an apparent cause for much rejoicing, the deposit being but a confused mixture of

[9] Sound information on Schieffelin and the location of Tombstone is scarce. This account has been synthesized from the typescript "History of the Discovery of Tombstone, Arizona, as Told by the Discoverer, Edward Schieffelin," and supplemented by secondary materials and local engineering tradition insofar as they seem consistent with the locale.

porphyry, flints, quartz, and quartzite. He grubbed about and found matters much improved a few feet down, where he hit the head of a steeply arched fold (anticline), one of whose strata was an iron-blackened limestone mineralized with silver much as in the manner of the Leadville ore. That location he staked as the Tough Nut (Schieffelin had a gift for sarcastic nomenclature). He found a fourth, the Lucky Cuss, and a fifth, which, because of a grubstake dispute and consequent lawsuit, was subsequently named the Contention. The final owners of the Contention may at times have wished that they had lost their suit, since later development revealed that the main ore shoot was pinched out five times in succession, thanks to the extreme faulting and displacement of the granitic country rock.[10]

The camp at Tombstone quickly blossomed forth. The region itself was singularly forlorn, but the rush much encouraged railroad building in the territory, and soon food, fuel, and whisky could be had at reasonable prices. The development was a good illustration of mining homiletics. The silver had been brought to the spot by a large porphyritic dike which had invaded and traversed the plateau. The dike was full of quartz, segregated by slow cooling, and the quartz in turn was full of large cubic pyrite crystals. Those had weathered out, leaving the quartz honeycombed with square cavities, many of which contained nests of hematite holding bits of native gold. The same weathering had taken out the silver and carried it down to limestone strata, where it formed a replacement deposit. The breakdown of the pyrites produced such a plentiful supply of ferric oxide that the miners went off shift so powdered with it that they looked like red men. The mines were worked by sinking a shaft on the ore from the top of the plateau. There the headframe and machinery were conveniently installed. When the mine was near the edge of the plateau, an adit for water, waste, and ore removal was run in to intersect the shaft, as in the manner of the Comstock's second stage.[11]

[10] Lakes, *Prospecting*, 240–44.

[11] The concentrators of the district were mostly at Charleston, a few miles away, where the San Pedro River water was abundant. Not much further treatment was attempted locally, owing to the cost of fuel. Contemporary photographs show that cordwood was used for the mine machinery, but it was far too expensive for smelting. At about this time began the practice of selling mine shares on a million-share basis, rather than on the traditional foot basis (Elliott, *Nevada's Twentieth-Century Mining Boom*, 17).

It was not long before the district came to resemble the Comstock in another fashion as well. Despite the elevation of Tombstone (nearly forty-five hundred feet) the water table is very high: a few miles away and a few hundred feet lower, the water virtually crops out at the startlingly lush little oasis of St. David. As the shafts of the Grand Central, Emerald, Silver Thread, and all the rest got down to this water table, the floods began in a manner that would have impressed Noah. Coordinated Cornish pumping by all the mines was the stop-gap solution, but when the day came that the levels went too low or too many of the mines got into the traditional water-table *borrasca* and quit or the price of silver dropped, Tombstone would dry up and blow away. Any one of the three disasters would be enough in itself. All three struck on virtually the same day in 1893.

Twelve years later an optimist named Frank Murphey organized the Development Company of America for the purpose of unwatering and reopening the Tombstone mines, somewhat after the manner of Sutro. He proposed to sink a master drainage shaft to the water table, install steam-driven force pumps at that point, and begin pumping out the entire district. When the water table had been lowered sufficiently, sinking in the master shaft would recommence, and the pumps would then be leapfrogged to the next lower level.[12] This process would be continued until a projected maximum depth of one thousand feet had been reached.

While dewatering was going on, the related Tombstone Consolidated Mining Company attended to reopening the mines as fast as they were dried out. By 1905 the project had proved a qualified success. At the eight-hundred-foot level the pumps were raising 2.3 million gallons of water daily, while the output of the reopened mines went to the refineries at El Paso in the form of two or three carloads of bulk concentrates a day. Profits were helped along by scavenging both low-grade ore and the waste dumps of the earlier period. With a rise in world silver prices that occurred at the same time, the operation showed a profit for four years.

In 1909 it was given out that boiler breakdown had shut down

[12] Elton W. Walker, "Sinking a Wet Shaft at Tombstone," *Mining and Scientific Press*, Vol. XCVIII (February 20, 1909), 284–86.

the drainage system and that before repairs could be effected the entire complex had been drowned beyond redemption. Counterarguments have been offered that the pumps themselves would not have been hurt or that the drifts would not have been seriously damaged by brief flooding; that a boiler failure on the surface could have been reasonably well repaired with speed; and that other devices could have been effectively called into play. For whatever reason seemed good to him, Murphey was apparently prepared to call it a day at Tombstone. The Phelps Dodge Corporation then showed some interest in the camp, but it too presently went away. As is invariably the case with deceased mining camps, there are a few Tombstone enthusiasts who will stoutly maintain that bonanza ore lies below in great quantities, just as there are individuals who will seek for the Lost Dutchman Mine until they die—it is differences of opinion that make mining history as well as horse races.

Matte Refining

Semifinal refining of matte pigs or mill concentrates was an ancient art and one which for the sake of clarity I have carefully avoided until this point. Some mining districts produced free gold, or matte so nearly free of extraneous elements that only minimal treatment was necessary before their bullion could be sold directly to the United States Mint, which always did its own final refining and adjustment of fineness. In California placers, for example, the only serious natural adulterants were iron and tellurium. The former was removed during amalgamation; the latter could be easily driven off as stack gas by low-heat roasting in the presence of oxygen. The gold-silver bullion of the Comstock Lode was almost equally free of everything unwanted but lead. Lead was dealt with by being oxidized to litharge, which floated to the surface of the melt and was readily skimmed off. On the other hand, the refractory ores of Colorado were composed of so great a variety of constituents that only smelting with a complex of fluxes could sort out the desirable metals, and often enough there had to be preliminary sortings by means of Wilfley tables or less effective devices in order to salvage values which would otherwise be dumped in the slags. It is a tolerably good generalization not only that each mining district displayed its own geological and extractive peculiarities but that each had as well its own peculiar ores, whose concentration and preliminary refining occupied the thoughts of many ingenious technicians.

Precious-metals refining was practiced so long ago as to be specifically mentioned in the Old Testament. One of the most ancient methods was cementation, which involved melting the matte, casting it into water to form beads, adding lead as a flux, and heating it carefully for a prolonged period in intimate contact with pulverized brick or potsherds. These earthy substances would absorb the lead and such unwanted metals as copper which the lead carried with it, leaving behind a fairly pure gold-silver button. This procedure was "trying in the fire," which gave reasonably satisfactory results when several times repeated, and certain cities such as Ionian Sardis had extensive refining facilities. The Palestinians and Ionians had to stop there until better methods were developed, as shown by the fact that the first known coinage was of "electrum," the gold-silver buttons that were the product of cementation —and of unaided nature upon occasion.

The Egyptians of the New Kingdom, however, had learned to separate gold from silver by melting electrum in a crucible at high heat, with salt and lead. The salt reacted with the silver and lead to form chlorides which could be skimmed off. This chloride slag could later be reduced by smelting to reclaim its metallic content. The question is why anyone would wish to do so, although tomb paintings of around 1000 B.C. show the process clearly. Air blast was furnished by foot-operated bellows, and the crucibles look surprisingly modern, as do the sticks or rods used to lift and pour out their contents. It seems probable that they desired reasonably pure gold for gilding and gold-leaf beating; gold leaf containing silver is too pale, while the presence of copper lends an unsightly greenish cast to the gold leaf. The process was expensive, but to a king whose afterlife was involved, economy did not matter. Electrum "coins," on the other hand, had a desirable degree of hardness, and until the time of Archimedes there was no ready way to determine whether the button had a bit less gold than it presumably contained.

The discovery of amalgamation led to the reverse process: removing the precious metals from the base elements. As before, the matte was beaten into thin sheets or dripped into water to form beads, which were hammered thin. The sheets or lenticular buttons were then immersed in mercury. The mercury amalgamated with the gold and silver, leaving the base-metal sponge behind. Amalgamation was by far the more expensive method, so the medieval alchemists and moneyers concentrated on improving the fire process, which could be used equally well for matte bullion or ore concentrates. In his *Pirotechnia* the Italian metallurgist Vannoccio Biringuccio, who wrote perhaps the first post-classical work on metalworking, antedating even Bauer, carefully describes the refining of concentrates in a manner which was hardly changed until the eve of the twentieth century:

First take by weight or measure the quantity of ore that you wish to smelt, [broken] into small pieces about the size of beans. If the ore had previously needed evaporation [roasting] by fire or cleaning [concentration] by washing, you will have had this done well and have had it properly prepared by a master sorter. A layer of this is arranged in front of the blast furnace, then a quarter part of galena or a third if you can obtain it. Then an equal part of crushed iron slag or slag from the same or different ore is added. These materials are spread in layers one above the other.

When the blast furnace is full of well-kindled burning charcoal, admit water to the bellows machine. When you see that it is well kindled and that the flames begin to come out forcefully on top, work it down with a little rake and fill it up again [with the ore and flux]. Add new charcoal and a basketful of the said composition of ore, proceeding in this way until the bottom of the blast furnace is filled with melted materials. Decide this by judgment, or by seeing through the mouth of the tuyere where the wind of the bellows enters that they are level with this. Then with an iron tool open the hole that was left as an outlet in front of the blast furnace and allow all the metal with its slag to run out. These two things, flowing through in the channel like an oil, enter the large forehearth into which all that is in the blast furnace is allowed to run out completely. Then, when the masters see that the forehearth is well filled, they plug up the hole of the blast furnace, put new materials in again, and continue to smelt.

The melted material that entered the forehearth separates by itself, the coarse and earthy parts [slag] remaining on top and the heavy ones sinking to the bottom. These earthy parts do not stand long in the air before they begin to harden. Then, with an iron fork strike it hard so that it is detached all around and floats, then place the fork under the corner and lift it out. In this way, as it cools, the slag is lifted out layer by layer until the clear metal is reached and no more slag can be seen on it.

> Now this metal that [remains] in the forehearth is of three natures: the two principal ones are of copper and lead and the third is of silver. The lead and the silver separate again from the more earthy nature of the copper and they go to the bottom while the copper remains above and begins to cool. Do with it just as you did with the slag, lifting it off layer by layer until you arrive at that leaden part. Then open the hole of the forehearth and let it [the lead-silver alloy] run into the ditch that is customarily made near by, and let it cool in this.[13]

Biringuccio's emphasis upon the removal of copper stemmed from the fact that it was the base metal which gave refiners the most trouble: it occurred frequently in both gold and silver ores, it was the favorite metal for either reasonable or deceptive alloying with both, it had nearly the same melting point as gold, and it was neither oxidized nor absorbed as readily as was lead.

Silver-lead bullion which was low in gold content was refined in a two-step process. Liquation, or what today would be called fractional crystallization, involved melting the bullion, then carefully reducing its temperature by slow degrees. At a critical point crystals of pure lead would begin to pop to the surface in obedience to the rule that alloys generally have lower melting points than that of either of their components. The lead crystals would be carefully skimmed off and put aside until it was apparent that no more were forthcoming. At that stage the bullion would have about five hundred ounces of silver to the ton and would be ready for cupellation. Reheating it and directing an air blast across the surface oxidized the lead to litharge, most of which could be skimmed off the top of the melt, and any remaining traces were absorbed in the bone ash or fired-marl cupel. That this cupellation was of respectable antiquity can be seen by reference to Jeremiah, who about 500 B.C. wrote, "The bellows blow fiercely: the lead is consumed of the fire."

Refining at the United States Mint

The final step in nineteenth-century refining was the prerogative of the United States Mint, which used the expensive but clean-cut acid process. Mill-run ingots purchased by assay value were re-assayed in the mint,

[13] Vannoccio Biringuccio, *The Pirotechnia of Vannoccio Biringuccio* (trans. and ed. by Cyril S. Smith and Martha T. Grudi), 153–54. The quotation has been slightly edited to omit redundancies.

then melted and subjected to cupellation to clean them of base metals. The gold-silver bullion remaining was then cast molten into water to reduce it to droplets, after which these beads were "parted" with nitric acid. The insoluble gold particles were filtered out, heat-annealed, and cast into bullion bricks each weighing about twenty-eight pounds. The silver was recovered from the solution as silver chloride and was reduced to metal by smelting. After 1875 the mint discontinued cupellation in favor of straight "wet" chemistry, which used the cheaper sulphuric acid to gain a somewhat more effective parting, a touch of acetic acid to get rid of unwanted traces of mill mercury, and hydraulic compression of the filtered gold in place of annealing.[14]

Bullion bricks destined for industrial use or for storage as currency reserves were assayed with extreme care and were stamped on one end with a number designating fineness; for example, a gold brick marked 9995 had five parts per ten thousand parts of base metal. Bullion destined for coinage was remelted; a carefully calculated amount of copper was added to bring it down to statutory fineness and desirable metallurgical characteristics; it was then rolled into blank sheets and transferred to the coinage department. Needless to say, worn or foreign coins as well as smelter dust and industrial scrap were treated in exactly the same fashion as was bullion fresh from the western mines.

Cyanidation

In the decade 1887 to 1897 there occurred a quiet but profound revolution in milling. It is surprising how little the contemporary accounts make of it, for the development of cyanidation made obsolete every mill and virtually every milling process used to extract gold. On the other hand, coming events cast their shadows before, and cyanidation was preceded by the semisolution which chlorination had appeared to offer to the problem of lost assay values. Millmen and chemists felt there ought to be some chemical process which could actually do what chlorination only promised—leach from tailing, waste, and low-grade ore virtually every atom of gold they carried. In Glasgow, Scotland, J. W. MacArthur, a metallurgical chemist, joined forces with the brothers

[14] Don Taxay, *The United States Mint and Coinage*, 87–89, 156–61. Sulphuric-acid parting of doré beads is supposed to have been developed about 1802 and then acquired by the Rothschild interests, who kept it a secret, to their great profit, for many years. See Russell H. Bennett, *Quest for Ore*, 168.

W. and R. W. Forrest, doctors of medicine, to seek in a dingy little laboratory the modern equivalent of the philosopher's stone. After great labor, they patented their answer to the problem. The Forrest-MacArthur process was based on the chemical fact that the cyanide radical is one of the very few common and reasonably cheap compounds for which gold has a decided affinity. Better still, the cyanides are not explosive, highly corrosive, or unstable. They are, to be sure, one of the deadliest of universal poisons, but when highly diluted they are not exceptionally dangerous. The opening act of the process was simplicity itself: merely run gold-bearing sands or slimes into a 0.005 sodium-cyanide solution, agitate well in the presence of air, and let nature take its course. Without fuss or trouble metallic gold would very quickly dissolve into the solution, which could then be decanted merely by opening a valve.

In practice, of course, there was much more to it than that. Free-milling gold ores responded very well after simple crushing and screening, as did placer sands: gold-to-gangue ratios of 1 to 100,000 were not too thin to be economically treated. More complex or refractory ores required thorough roasting, preliminary chemical treatment, and considerable adjustment of the cyanide solution or time of contact, but withal little that was beyond the power of a competent chemical engineer to devise. Recovery of the dissolved gold was equally simple. The pregnant liquor was run through beds or boxes of fine zinc shavings, which precipitated the gold back to metallic form in exactly the same manner as iron reduced copper sulphate to metal in sulphate leaching. In most cases the gold came out of solution in amorphous form as a black sludge, contaminated somewhat with undissolved metallic zinc. In other cases it adhered to the zinc shavings, but in both instances refining was elementary, since a peculiar virtue of zinc is that it will volatilize just as mercury does at the right degree of heat, and can be retorted off in virtually the same manner. The resulting sponge can be cast and sold directly to the mint without further ado.[15]

The cyanide process was first used on a production scale at the Crown Mine at Karangahake, New Zealand, in 1889. Two years later it was tried in the United States. In 1889 cyanide extracted only six thousand

[15] Bray, *Nonferrous Production Metallurgy*, 252–63. For an account of cyanidation in Mexico, see Rickard, *Journeys of Observation*, 54–68.

dollars in gold; two years later the figure was six million dollars, and virtually every precious-metal mill in the world was converting to cyanidation. They kept their crushers and stamps but swept away the pans, mullers, separators, and tables to make room for the big wooden vats. Then the stamps went, too, replaced by ball mills whose fine-crushing ability was now an asset instead of a liability. The steel cyanide drum in which shipment was made, a product of I. G. Farbenindustrie, A. G., became nearly as common a sight in the mining camps as the empty wooden dynamite case.[16]

The Close of the Mining Frontier

Cyanidation set off a gold rush contemporary with, perhaps greater than, but infinitely less well known than the Yukon or Witwatersrand booms. The cyaniders descended upon every old tailing heap in the world, ran it through their vats, and decamped with the proceeds. Ten thousand low-grade gold shoots suddenly became profitable to mine, with cyanide making it economical to mill the ore cheaply right next to the shaft collar. But this boom attracted little attention because it was devoid of what the public considers glamor. There was no prospecting except by attorneys among the dusty patents of defunct mining companies. There was no bustle or wasteful scurrying to and fro—at most, there were small steam shovels tossing sands into ball mills, which in turn spilled slimes into a few redwood vats from which clear light-green liquid was run into zinc boxes at infrequent intervals. Free-lance operators could hand-shovel tailing into fifty-gallon drums to the same effect. The process water was used over and over again. From time to time unattractive sludges or shavings were shipped off to refineries at distant points. It was a very dull business, and it did little to stimulate moribund mining camps since it employed so few men. All it did was double the world's annual production of gold. There were, to be sure, big strikes and feverish boom camps after 1893—Goldfield and Tonopah, Skagway, Searchlight, and Creede—but somehow things were not the same.

[16] German domination of mining machinery outside the United States was virtually absolute between 1890 and 1914. I. G. Farben, which had a virtual monopoly of the chemical industry, produced the cyanide, while the Krupp Works produced the most popular line of milling equipment.

Cyanide rang down the curtain upon the frontier period of American gold and silver mining, although cyanide was really more a symptom than the cause. Far more constructively important in closing out three thousand years of technology was the presence in the twentieth-century mining camps of the young university-trained engineers in high-laced boots and choke-bore breeches, careening about in Model T's. Their business was, in the words of one, "to do for sixpence what any fool could do for a shilling," and, to the amazement of grizzled Cornishmen and Irishmen, they did just that. They supplanted the desert rat as the prime mover of mineral-district culture, while their young wives organized clubs to build schools, close down the hog ranches, and make the menfolk leave their revolvers locked up where the children could not get at them. Though no one realized it at the time, the old mining frontier had closed.

Its passing left more than the waste dumps, ghost towns, and rotting mine timbers which are its monuments. The frontier unquestionably hastened the settlement of the West, drawing there a population of comparatively good education and settled habits. Its requirements speeded the growth of agriculture, of communication, and of all the arts. The mining excitements promoted education and the law. Its real leadership rested with the great consulting engineers, but it has often escaped notice that these men were much more than technicians. They were scholars, molders of public opinion, and personal leaders whose influence was invariably exercised for the better. If they were as proud as Lucifers, they had good reason for their pride. A Deidesheimer or a Rickard did more for the West than did a thousand gunmen like Wild Bill Hickok or Wyatt Earp.

It was the West's good luck that it was never troubled by such friction as that which marred the history of South Africa in similar circumstances. The Boer-British troubles seem to have been based on the same issues as those that rose between forty-niners and the Mormons, but in the American West irritation did not go beyond hard looks and carping words. Perhaps the comparative sophistication and the high discipline of the Latter-Day Saints was the decisive element, in contrast to the virtual anarchy of the Boers, who had, moreover, been virtually isolated from Europe for 250 years before the Witwatersrand rush. The closest the Americans came to such a clash was in the matter of

Chinese labor. It was averted when interested parties closed ranks determined to resist by force if necessary even the suggestion of Chinese workers. It was undoubtedly hard on the Chinese immigrants, but it would have been a great deal harder on the Cornish and Irish majority

Fig. 82. Early hoisting. Reproduced by permission of Buck O'Donnell and Shaft and Development Machines, Inc.

if unscrupulous highbinders such as William Sharon had been allowed to introduce Chinese labor to the mines.

Economically speaking, it would seem that in the long run, despite its image as a bonanza frontier, the mining West scarcely broke even. Costs insatiably devoured profits, and the number of men and institutions who came away with substantial gains was remarkably small. Most of the mineral strikes were tempting at first. They attracted millions of dollars (I decline even to speculate how many million) from Europe and the Northeast, but returned comparatively few in dividends. Nevertheless, the money was not wasted. It was expended in payrolls, transportation, cost of materials, and construction of which most was locally purveyed and from which the general locality benefited. No matter what tenor of ore it was hoisting, a working mine was doing its share toward western development.

Appendices

APPENDIX I

CONCENTRATION AND MILLING FLOW CHART

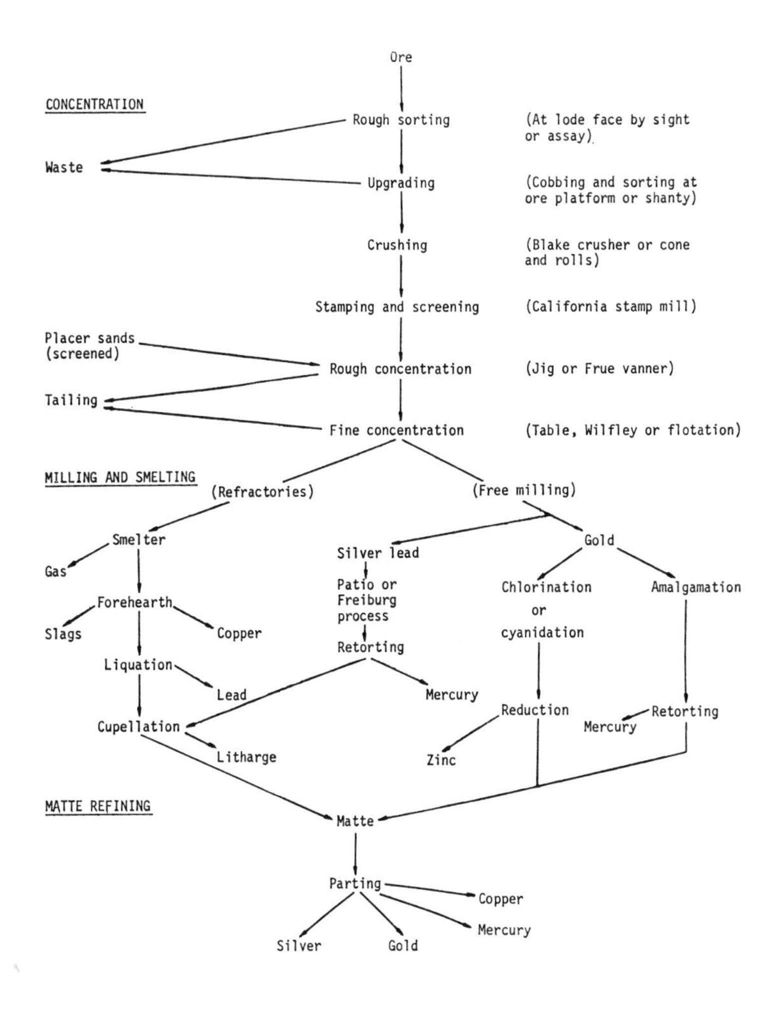

GLOSSARY OF MINING TERMS in Common Use in the Mining States and Territories[1]

The Cornish and Spanish or Mexican terms have been selected, because they are those chiefly used. In the Nevada, Colorado, Utah, and California mines the Cornish terms have become common, owing to the large number of workmen from Cornwall who have been and are employed. In Arizona and New Mexico, as well as in southern Colorado, the Mexican terms are in common use, because Mexicans are the principal workmen.

Cornwall Mining Terms

A.

Acicular—Slender and straight crystals.

Adit level—A horizontal excavation through which the water drawn from the bottoms of the mine thereto by the engine, and that from above, passes off to the surface. This level is usually commenced from the bottom of the deepest neighboring vale, and extended throughout a great part of the mine.

Aggregated—Where the component parts only adhere together, and may be separated by mechanical means.

Air-machine—An apparatus for forcing fresh air into, or withdrawing foul air from, badly ventilated places.

Air-pipes—Tubes or pipes of iron or wood, for ventilating underground, or for the conveyance of fresh air into levels having but one communication with the atmosphere, and consequently no current of air.

Aitch-piece—That part of a plunger-lift in which the clacks are fixed.

Alliaceous—The garlic odor of arsenical minerals when heated or struck.

Amorphous—Without form.

Anhydrous—Without water of crystallization.

Arborescent—Ramifying like a tree.

Arch—A piece of ground left unworked near a shaft.

Arched—The roads in a mine, when built with stones or bricks, are generally arched level drifts.

Argillaceous—Consisting of clay.

Arsenicate—The arsenic acid united with a base, as copper in the arsenicate of copper.

Ante—Rubbish, containing little or no ore.

Average produce—The quantity of fine copper contained in 100 parts of ore; thus, a parcel of ore, having a produce of 10⅝, contains 10⅝ *per cent* of copper, being above the average of copper ores in Cornwall.

Average standard—The price per ton of fine copper in the ore, after adding

[1] From Richard J. Hinton, *The Handbook to Arizona: Its Resources, History, Towns, Mines, Ruins and Scenery*, San Francisco, 1878.

returning charges for smelting, of £2 15s. per ton of ore, in Cornwall, and £2 5s. per ton of ore at Swansea.

Axis of a crystal—The lateral planes surround its axis, which is an imaginary line passing down the middle of the prism from the center of the upper to that of the lower terminal plane.

B.

Back—the back of a lode is the part nearest the surface. The back of a level is that portion extending above it to within a small distance of the level next above.

Bal—The miners' term for a mine.

Bar of ground—A vein of different description of rock, etc., from that in its vicinity.

Base—The substance to which an acid is united.

Batch of ores—Certain quantity of ore sent to the surface by any *pare* of men.

Bearers—Supports to the pump in the engine shaft.

Beat away—To excavate; usually applied to hard ground.

Bed—A seam, or horizontal vein of ore.

Bend—Indurated clay; a name given by miners to any indurated argillaceous substance.

Bit—The steeled end of a borer.

Black-jack—Blende.

Black tin—Tin ore ready for smelting.

Blast—The air introduced into a furnace.

Blast holes—The holes through which the water enters the "windbore," or bottom of a pump.

Blasting—Forcing off portions of rock by means of gunpowder. A hole is made with a borer, into which gunpowder is inserted, then confined, and set fire to.

Blende—One of the ores of zinc, composed of iron, zinc, sulphur, silex and water; on being scratched it emits a phosphoric light.

Block tin—Metallic tin.

Blower—A smelter.

Bob—The engine beam.

Borer—A boring instrument, with a piece of steel at the end, called a boring bit.

Botryoidal—Globular forms such as are found in copper, etc.

Bottoms—The lowest workings either in a slope, level, or elsewhere.

Boulders—Large stones or pebbles.

Bounds—The proprietary of tin ore over a given tract.

Bonney—A distinct bed of ore that communicates with no vein of ore.

Brace—The platform placed over the mouth of a shaft, or winze, and to which the tackle is fixed.

Branch—A small vein which separates from the lode, and very generally unites again therewith.

Brood—Impurities mixed with the ore.

Bryle—The traces of the presence of a lode, found in the loose matter, on or near the surface.

Buckers—Bruisers of the ore.

Bucket—The piston of the lifting pump.

Bucket-lift—A set of iron pipes attached to a lifting-pump.

Bucket rods—Wooden rods to which the piston of a lifting-pump is attached.

Bucking iron—The iron, or tool, with which the ore is pulverized.

Bucking plate—An iron plate on which the ore is placed for being bucked.

Buddle—An apparatus by which the stamped tin is washed from its impurities; there are various contrivances in use; Brunton's Frame, the Round Buddle, and Zenner's Rotating Buddle, being the most approved.

Buddling—Separating the ores from the earthy substances, by means of an inclined hatch, or cistern.

Bunch or squat of ore—A quantity of ore of small extent; more than a "stone," and not so much as a "course."

Bunney—See *Bonney.*

Burden or oresburn—The substances reposing on a bed of stream tin ore.

Burning house—The furnace in which ores are calcined to sublime the sulphur from pyrites; the latter, being decomposed, are more readily removed by washing.

Burrow—A heap of deads, attle and rubbish.

C.

Cage of a whim—The barrel on which the rope is wound up.

Cal—Wolfram.

Callys—See *Killas.*

Caud or Kaud—Flour.

Capel—A stone composed of quartz, schorl and horneblende, usually occurring on one or both walls of a lode, and more frequently accompanying tin than ores.

Captain—One of the superintendents of the mine.

Captain dresser—Superintendents of the dressing of ores.

Carrack—See *Capel.*

Cast after cast—The throwing up of tin stuff, etc., from one stage of boards to another; the stages are placed about a fathom apart.

Cases of spar—Veins of quartz (not containing ores) which have not a direction parallel to the lodes.

Casing—A division of wood planks, separating a footway or a whim, or an engine shaft, from one another.

Cathead—A smaller capstan.

Cannter lode—A lode which inclines at a considerable angle with the direction of the other bodies in the vicinity.

Charger—An implement in the form of the bit of a carpenter's auger, for charging holes or blasting, which are dug horizontally.

Chats—Small heaps of ore.

Chimming—A process of similar effect to tossing, but being performed on small quantities of ore, the keeve is supported on the verge of its bottom.

Clack—The valve of a pump of any description.

Clack-door—The aperture through which the clack of a pump is fixed and removed.

Claying—Lining the hole (in which gunpowder is to be placed) with clay, to prevent the powder becoming damp.

Cob—To break the ore with hammers in such a manner as to separate the dead or worthless parts.

Cockle—Schorl.

Cofering—Securing the shaft from the influx of water by ramming in clay.

Coffin—Old workings open to the day.

Collar of a shaft—The timber by which its upper parts are kept from falling together.

Collar launder—The pipe or gutter at the top of a lift of pumps, through which the water is conveyed to the cistern.

Connection, or *connecting rods*—The larger rods which are attached to the engine beam.

Core—Miners usually work but six hours at a time, and consequently four pairs of men are required for the whole time—"forenoon core," from 6 A.M. to noon; "afternoon core," from noon to 6 P.M.; "first core by night," from 6 P.M. to midnight; and "last core by night," from midnight to 6 A.M.

Costeaning—Discovering lodes by sinking pits in their vicinity, and driving transversely in their supposed direction.

Country—The strata or rock through which the vein or lode traverses.

Course of ore—A portion of the lode containing a regular vein of ore.

Cover—The box into which the ore is removed from the rock; also the place at the head of the trunk, in which the stimes are, by agitation, mechanically suspended in water, in the process of trunking.

Creases—Divisions of buddled work.

Crib, or Curb—A circular frame of wood, screwed together, as a foundation for bucking, or pulverizing ore in a shaft.

Crop—The best ore.

Cross-course—A lode, or vein, which intersects or crosses a lode at various angles, and generally throws the lode out of the regular course.

Cross-course spar—Radiated quartz.

Cross-cut—A level driven at right angles to the direction of the lode.

Crushing—Grinding the ores without water.

Cube—A solid figure contained under six equal sides.

Cuneiform—Wedge-shaped.

Cupelo—A small furnace.

Cut—To intersect by driving, sinking, or rising.

D.

Dam—Choke damp, foul air.

Deads—Attle or rubbish.

Dead ground—A portion of the lode in which there is no ore.

Dean—The end of a level or cross-cut.

Dialling—Surveying for the purpose of planning.

Dileuing or Terluing—Washing ores supported on a hair-bottomed sieve in water.

Dippa—A small pit.

Dish—That portion of the produce of a mine which is paid to the landowner or lord.

Dissuing—Is when the lode is small and rich, to break down the strata from one of its walls, by which means it can afterwards be taken away without being deteriorated and without waste.

Dowsing rod—The hazel rod of divination, by which some pretend to discover lodes.

Draft-engine—An engine used for pumping.

Dredging-ore (also called *dradgyore* or *drady-trade*)—A stone impregnated or traversed by minute veins of ore.

Dressers—Cleaners of the ore.

Drift—The excavation made for a road underground.

Driving—Digging horizontally.

Dropper—A branch when it leaves the main lode.

Dry—A place fitted witih steam pipes and other heating apparatus, wherein miners' underground garments are dried.

Durns—A frame of timber with boards placed behind it, to keep open the ground in shafts, levels, etc.

Dzhu, or Hulk—To dig away a portion of the rock, etc., on one side of the end, that the blast may be more efficient.

E.

Elvan—Porphyry, clay, stone.

End—The further extent of a level or cross-cut.

Engineer—The superintendent of the machinery.

Engine man—Man who attends to and works the engine.

Engine-shaft—The pit or shaft by which the water is drawn by the engine from the lower parts of the mine to the adit, or surface.

F.

Fang—A niche cut in the side of an adit, or shaft, to serve as an air-course; sometimes a main of wood pipes is denominated a fanging.

Farm—That part of the lord's fee which

is taken for liberty to work in tin mines only that are bounded—generally one-fifteenth.

Fast—The firm rock beneath the diluvium.

Feeder—A branch when it falls into the lode.

Flang—A two-pointed pick.

Flat-rods—Rods for communicating motion from the engine horizontally.

Floran tin—Tin ore scarcely perceptible in the stone; tin ore stamped very small.—*Pryce.*

Flookan—A soft, clayey substance, which is generally found to accompany the crop courses and slides, and occasionally the lodes themselves; but when applied to a vein, means a cross-vein or course composed of clay.

Fluke—The head of the charger; an instrument used for cleansing the hole previously to blasting.

Foot-wall—Is the wall under the lode; it is sometimes also called the underlaying wall.

Foot-way—The ladders by which the workmen ascend and descend.

Force-piece—A piece of timber put in a level, shaft, etc., in a diagonal position, for keeping the ground open.

Fork—"Water in fork," water all drawn out; the bottom of the engine shaft.

Furnace—The place in which the ore is placed for the purpose of smelting or reduction.

G.

Gad—A pointed wedge of a peculiar form, having its sides of a parabolic figure.

Glist—Mica.

Good levels—Levels driven nearly horizontal.

Gossan—Oxide of iron and quartz, generally occurring in lodes at shallow depths.

Grass—The surface.

Grain tin—Crystalline tin ore; metallic tin, smelted with charcoal.

Grate—Stamps grate; a metallic plate pierced with small holes; it is attached to the stamps, and through the holes the stamped ores escape.

Griddle or Riddle—A sieve.

Grinder—Machinery for crushing the ores between iron cylinders or barrels.

Ground—The country; the stratum in which the lode is found.

Growan—Decomposed granite; but sometimes applied to the solid rock.

Guag—A place that has been wrought before for tin.

Gulph of ore—A very large deposit of ore in a lode.

Gunnies—Levels or workings.

Gurt—A gutter; a channel for water.

H.

Halvonner—The dresser of, or operator on, the halvans.

Halvans—The ores which are not sufficiently rich to be offered for sale until much of the impurities with which they are mixed are removed by operations in water.

Hanging wall—The wall or side over the lode.

Hauling—Drawing ore or attle out of the mines.

Head-sword—The water running through the adit.

Head tin—A preparation of tin ore towards the working it into metal.

Heave—The horizontal dislocation which occurs when one lode is intersected by another having a different direction. A right or left hand heave is when the part of the intersected lode on the opposite side of the traversing vein is found by turning either to the right or the left.

Hoggan—the tinner's pasty.

Hook-handles—The handles by which a windlass is worked.

Horse—The dead ground included between the branches of a lode, at the point of their separation.

House of water—A vugh or space, whether artificially excavated or not, filled with water.

Horse arm—The part of the horse whim to which the horses are all attached.

H piece—See *Aitch piece.*

Huel—See *Wheal.*

Hutch—Cistern or box.

I.

Irestone—Hard clay slate, hornblende, hornblende slate, hornstone.

J.

Jigger—Cleaner of ores.

Jugging—Separating the ore with a griddle, or wire-bottomed sieve; the heavier substance passing through the bottom, or lower part of the sieve; the lighter substances remaining on the upper part are put by for halvans.

Junction—Applied to where veins unite.

Jumper—A long borer worked by one person.

K.

Keeve—A large vat.

Kibble—A bucket, usually made of iron, in which the ores, etc., are drawn to the surface.

Kibble filler—Man who sends up work, etc., to the surface.

Killas—Clay slate.

L.

Lander—Man who attends at mouth of shaft to receive the kibble in which ores, rubbish, etc., are brought to the surface.

Lappior—The dresser of the leavings.

Laths—The boards which are put behind and supported by the "durns."

Launders—Tubes or gutters for the conveyance of water; their form that of a long box wanting the upper side and both ends.

Lead spar—Sulphate of lead.

Leader of the lode—A branch or small vein; part of the main lode.

Learies—Empty places; old workings or vughs.

Leat—A water-course.—*Pryce.*

Leavings—The ores which are left after the "crop" is taken out.

Levels—Galleries driven on the lode, usually at ten, twenty, thirty, etc., fathoms below the surface.

Lifters—Wood beams to which the iron heads of a stamping mill are fastened.

Lock-piece—A piece of timber used in supporting the workings.

Lode—A regular vein producing or affording any kind of metal.

Lode storran—A drang driven towards rising ground on the indications of a lode in marshy ground.

Loobs—Slime containing ore.

Lost levels—Levels which are not driven horizontally.

M.

Machine whim—A rotary steam engine employed for winding.

Mallet—An instrument used with the borer.

Material man—One who delivers out and has care of the materials.

Meat earth—The vegetable mold.

Mock lead—Blende.

More—A quantity of ore in a particular part of a lode, as a "more" of tin.

Moorstone—Granite.

Mundice—Iron pyrites.

Mun—Any fusible metal.

N.

Needle, or Nail—A long taper piece of copper or iron, with a copper point, used when stamping the hole for blasting, to make, by its withdrawal, an aperture for the insertion of the rush or train.

Nogs, or Nays—Supports for the roof of a mine.

P.

Pack—To occasion the speedy subsidence of the ore in the process of tossing or chimming, by beating the keeve in which it is performed by a hammer.

Parcel—A heap of ore dressed and ready for sale.

Pare—Gang or party of men.

Pass—An opening left for letting down stuff to the levels.

Peach—Chlorite.

Pedn cairn—A bunch of ore at a distance from the lode.

Pick—An instrument in common use, as well in agriculture as in mining.

Picker, or Poker—A hard chisel for dzhuing, which is held in one hand and struck with a hammer.

Pillar—A piece of ground left to support the roof or hanging wall.

Pitch—Limits of the piece of ground set to tributers.

Pitch bag—A bag covered with pitch, into which powder is put (previously to its being introduced into a damp hole) that it may be protected from moisture.

Pitman—One employed to look after the lifts of pumps and the drainage.

Pitwork—The pumps and other apparatus of the engine shaft.

Plat—An excavation or place to contain any ore or deads.

Plunger—The piston or forcer of a forcing pump.

Plunger lift—The set of pipes attached to a forcing pump.

Point of the horse—The spot where the vein is divided into two or more branches.

Pol-voz (pronounced polrose)—The pit underneath a water-wheel.

Pot-growan—Soft decomposed granite.

Prian—Soft white clay, esteemed a favorable sign when found in a lode.

Pricker—A thin piece of iron used to make a hole for the fuse or match to fire a blast.

Prit—A solid piece of virgin metal, or the button from an assay.

Produce—Fine copper contained in 100 parts of ore.

Purser—The cashier or paymaster at the mines.

Q.

Quere (also spelled*queere* and *queear*) —A small cavity or fissure.

R.

Rack—An inclined plane on which the ores and slime are washed and separated.

Racking—Is a process of separating small ore from the earthy particles by means of an inclined wooden frame; the impurities being washed off, and the ore remaining near the head of the rack, taken from thence, undergoes tossing.

Reed, or *Spire*—Gorse, or other tubular vegetable, into which gunpowder is put to convey a train from the snoff to the charge, the reed being put into the aperture made by the needle.

Refining—Separating the ores.

Relief—When one workman of the same pare changes core, or takes the place of another.

Riddle, or *Griddle*—A sieve.

Rising—Digging upwards.

Row—Large stones, rough.

Rullers—The persons who work the barrows under ground.

Run—When excavations fall together.

Run of a lode—Its direction.

Rush—Used for the same purpose as the reed and spire.

S.

Scal—A shale, or portion of earth, rock, etc., which separates and falls from the main body.

Scovan lode—A lode having no gossan on its back, or near the surface.

Scraper—A piece of iron used to take out the pulverized matter which remains in the hole when bored, previously to blasting.

Seam—A horse load.

Sett—A lease, the boundaries and terms of mining ground taken by the adventurers.

Set of timber—A frame complete to support each side of the vein, level, or shaft.

Set-off—The part of a connecting-rod to which the bucket-rod is attached.

Shaft—A sinking or pit either on the lode or through the country.

Shaking—Washing the ores.

Shammel—When ore or water is lifted part of the required height by one machine or person, and part by another.

Shears—Two very high pieces of wood placed in a vertical position on each side of a shaft and united at the top, over which, by means of a pulley, passes the capstan rope. This is for the convenience of lifting out or lowering into the shaft timber or other things of great length.

Shelf—The firm rock.

Shieve—The pulley over which the whim rope passes.

Shoding—Tracing round stones from the vale to the lode whence they were torn.

Shooting—Shutting or blasting; fracturing and separating by the use of gunpowder.

Sinking—Digging downwards.

Skimpings—Skimmings of the light ores in the dressing process.

Slide—A vein of clay, which, intersecting a lode, occasions a vertical dislocation.

Slimes—Mud containing metallic ores; mud or earthy particles mixed with the ore.

Smelting—Reducing the ore by means of fire.

Snoff, or *Match*—A substance, frequently brown paper, or other slowly combustible substance, which is ignited at one end, the other being in contact with the rush or train in blasting; the slow combustion is to permit the escape of the laborers.

Sollar—A small platform at the end of a certain number of ladders.

Spalling—The breaking up into small pieces, for the sake of easily separating the ore from the rock, after which it undergoes the process of cobbing.

Span beam—The horizontal beam passing over the whim in which the upper pivot of the perpendicular axis moves.

Stamps—Machinery for crushing the ores with the presence of water.

Stamp's head—The iron weight or head connected with the stamps.

Standard—The price of fine copper.

Stem—A day's work.

Stope—A horizontal bed; or ground adjacent to the levels; *to stope*, to excavate horizontally, layer after layer.

Spar—Quartz.

Spend—To break ground; to work away.

Squat of ore—See *Bunch*.

Strake—A launder, or box of wood without ends, in which the process of washing or tying is performed.

Strapping plates—The iron plates by which the connection rods are fastened to each other.

Stream tin—Tin ore found in the form of pebbles, most frequently in vales.

Streamers—The persons who work in search of stream tin.

String—A small vein.

Stuff—Attle or rubbish.

Stull—Timber placed in the backs of levels, and covered with boards or small poles to support rubbish.

Sturt—When a tributer takes a pitch at a high tribute, and cuts a course of ore, he sometimes gets two, three to five hundred pounds in two months; this great wages is called a sturt.

Sump—A pit; the bottom of the engine shaft.

Sump-shaft—The engine shaft.

Sump men—Men who assist the pitmen, sink the engine shaft, and attend to the machinery in the engine shaft.

T.

Tackle—Windlass, rope, and kibble.

Tamping—The material, usually soft stone, placed on the gunpowder, in order to confine its force, which could otherwise pass up the hole; also, the process of placing the material.

Tamping iron or bar—Tool used for beating down the earth substance on the charge used in blasting.

Team—To lade water in bowls.

Thrown (either up or down)—Is when a slide intersects a lode, the dislocation being shown by a transverse section. *Thrown up*, is when the undiscovered portion of the intersected lode is found to have been apparently lengthened. *Thrown down* is the reverse.

Ticketings—The sale of ore.

Timber man—The man employed in placing supports of timber in the interior of the mine.

Tollar—A person who periodically examines the limits of ground producing tin ore belonging to himself, or (the lord) his employer.

Tossing, or *Tozing*—A process consisting in suspending the ores by violent agitation in water, their subsidence being accelerated by packing; the lighter and more worthless matter remains uppermost.

Trade—Attle or rubbish.

Tram carriage—The carriage (usually made of iron) used on a tram road.

Tram-road—Iron railroad way.

Treloobing—See *Tossing*.

Tribute—Proportion of the ore which the workman (tributer) has for his labor.

Tributers—Men whose pay is a certain portion of the ore, or value of the ore they raise.

Tribute pitiches—The limited portion of a lode which is set to "a pare" of tributers, beyond which they are not, for the time being, permitted to work.

Trunk—A long narrow cistern or pit, in which the ore and slimers which are mixed are separated by the subsidence of the former, and the washing off the impurities; the inclined box in which the ore and slimy impurities are separated in the process of trunking.

Trunking—Process of extracting ores from the slimes; subsequently the ores undergo the process of racking and tossing.

Tummals—A great quantity; a heap.

Tunnel head—The top of a furnace, at which the materials are put in.

Turning house—The first cutting on from the slimes; subsequently the the lode after it is cut in a cross-cut.

Turned house—A term used when a level, in following branches of ore, is turned out of the original direction.

Tutwork—Work in which the laborer earns in proportion to the amount of his labor, being paid for driving at a certain price per fathom.

Tuyere—The aperture through which the air or blast is introduced into the furnace.

Tying—Washing.

U.

Underlayer—A perpendicular shaft, sunk to cut the lode at any required depth.

Underlay-shaft—Shaft sunk on the course of the lode.

V.

Van—To wash and cleanse a small portion of ore on a shovel.

Vugh, Vugg, or *Vogle*—A cavity.

W.

Washing—The ore undergoes occasionally two or three washings; the first process being that of washing the slime and earthy particles from the rougher and larger stones of ore.

Water in fork—When all the water is extracted.

Well—The lower part of a furnace into which the metal falls.

Whim—A machine worked by horse, steam or water, for raising ores, etc.

Whim driver—Man who attends to the horse in the whim.

Whim rope or *chain*—The rope or chain by which the kibble is attached to the winding engine or whim.

Whim shaft—The shaft by which the stuff is drawn out of the mine by horse or steam whim.

Whip and derry—A kibble drawn to the surface by a horse, the rope attaching one to the other simply passing over a pulley.

Willde lead—Blende, zinc ore.

Winch or Winze (contraction of windlass)—The wheel and axle frequently used for drawing water, etc., in a kibble by a rope.

Windbore—The lowest pump, in which there are holes to admit water.

Winding engine—One used to draw up ore, attle, etc.

Winze—A sinking on the lode communicating one level with another, for proving the lode, or for ventilating such drivings.

Work—Ores before they are cleansed or dressed.

Working barrel—The pump in which a piston works.

Working big—Sufficiently large for a man to work in.

Z.

Zawn—A cavern.

Zighyr—When a small, slow stream of water issues through a cranny it is said to zighyr or sigger.—*Pryce.*

Spanish Mining Terms

A.

Abra—A fissure; a considerable opening or cavity in the mountain, rock, or lode.

Abronzado—Yellow copper ore, sulphuret of copper.

Acarreadores—Wood-carriers.

Acero—Steel.

Achicar—To decrease or diminish; applied to the diminution of water in any of the workings, lowering the water in the shafts, etc.

Achicadores—Workmen employed in removing the water in *botas.*

Acuña—Die for coining.

Acuñacion—Coining.

Acuñador—One who coins.

Acuñar—To coin.

Ademador—A mining carpenter; a timber man.

Ademar—To timber.

Ademe—Timber work for securing and supporting the works of the mine.

Adobes—Unburnt bricks made of straw, adhesive earth, and dung dried in the sun.

Administrador—Superintendent.

Administracion—Management.

Afinacion—Refining.

Agata—Agate.

Agua fuerte—Aquafortis, nitrous or nitrous acid.

Ahondar—To sink, to deepen.
Ahonde—Sinking or driving downwards.
Alabastro—Alabaster.
Atbañil—Mason, bricklayer.
Albaredon—A dyke.
Albayalde—White lead.
Albergue—A natural hollow, a den.
Alcohol—Galena, sulphuret of lead, antimony.
Alcribis o Tovera—The tuyere of a smelting furnace.
Alear—To alloy metals.
Aleacion—The art of alloying metals.
Althondiga—Corn market or public granary.
Alimentos—In mining, an allowance to mine owners, as subsistence, until their mines become profitable.
Almacen—A store-house, store-room, warehouse.
Almadeneta—A stamp head.
Almagra—Ruddle, red ochre.
Almud—Twelfth part of a fanega.
Alquifal—Galena.
Alquilar—To hire.
Alta—The upper part.
Alumbre—Alum.
Ambar—Amber.
Amatista—Amethyst.
Amianto—Amianthus.
Amoldar—To mold.
Amonedar—To coin.
Amparo—The maintenance of the legal right of ownership by continued possession. In mining this can only be preserved by keeping a certain number of men at work at certain periods, as determined by the mining code.
Anchura—Width, roominess.
Angulo—An angle, a corner.
Antimonio—Antimony.
Apareje—A table, a block and fall, an apparatus; a set of harness for beasts of burden or draft; a pack-saddle.
Apartado—Establishment for parting silver and gold.
Aperos—Utensils; also materials, such as gunpowder and paper for blasting, etc.
A pique, Trabajar a pique—Digging downwards in a vertical direction.
Apolvillados—Rich ores.
Apuradores—Men who re-wash the earth from the tinas.
Arcilla—Clay.
Arena—Sand.
Arenilla—Fine sand.
Arrastrar—Applied to where veins unite and form one; to drag.
Arrastra—Mill for grinding ores, employed in the process of amalgamation of silver ores and of gold; a crushing mill.
Arreador—Horse-driver for malacates.
Arroba—25 lbs. Spanish weight.
Arriero—A muleteer.
Arsenico—Arsenic.
Asbesto—Asbestos.
Asserrador—A sawyer.
Aserrar—To saw.
Asfalto—Asphaltum.
Astillero—Open forest, pasture for mules, etc.
Atacadero—A rammer.
Atacador—Rod for ramming in the charges for blasting.
Atajador—A boy who attends the horses and mules.
Atajo abierto—Applied to a mine when worked in the manner of a quarry, or by an open cut in a rock or mountain.
Atargea—Water-course of masonry.
Atecas—Laborers who collect the water in buckets from the planes of the mines, in order to pass it off by the shafts; also, men who fill the skins in the shafts with water, mud, etc.
Atierras—Attle, rubbish; in the mine, earth preventing the continuation of the work.
Atisador—A stoker; man who attends the furnace.
Audiencia—Principal tribunal of justice.

Aviado—The mine-owner supplied with funds for working his mines.
Aviador—He who supplies funds for working mines.
Avio—Funds advanced for working mines.
Avios—Implements.
Ayudante—Assistant.
Azabache—Jet.
Azanca—A leat.
Azarcon—Red lead.
Azogue—Quicksilver; silver ore adapted for amalgamation.
Azogue apolvillado—Very good ore for amalgamation.
Azogue comun—Common ore for amalgamation.
Azogue ordinario—Ordinary ore for amalgamation.
Azogue razonable—Middling ore for amalgamation.
Azogue en caldo—Quicksilver.
Azogueria—The wareroom in which quicksilver is kept in store.
Azoguero—An amalgamator; person who superintends the process of amalgamation.
Azufre—Sulphur.
Azufre vivo—Native sulphur.

B.

Bancos—Rocks which intercept the vein, or cause it to take a different direction.
Banquillos—Stools on which the marquetas are placed.
Baño—The last portion of quicksilver applied to a torta.
Barquina—A large furnace.
Barquines—Forge bellows.
Barra—A bar, an iron crow; equal shares into which the interest in a mine is divided, usually twenty-four in number.
Barra de plata—A bar of silver, usually about 135 marcs, or 1080 ounces.
Barranca—A ravine.
Barrena—A drill or borer used in blasting.
Barrenadores—Miners who work with the borer and mallet.
Barrenar—To bore.
Barrenero—A boy who attends with the boring tools.
Barrenos—Holes made for blasting.
Barreta—A miner's bar or crow.
Barreteros—Miners who work with crowbars, wedges and picks.
Barro—Clay, loam, mud.
Basalto—Basalt.
Batea, or *Apuradera*—A bowl used in rewashing.
Bajo—Beneath, lower part.
Beneficiar—To extract the metal from the ore; to dress ore.
Beneficio—Making the metallic contents of the ore available by reduction.
Beneficio de cazo—Reduction of ore by amalgamation, conducted in a copper pan over a fire: a hot amalgamation.
Beneficio de hierro—Reduction of ore by amalgamation, with the addition of fragments of iron.
Beneficio de patio—Reduction of ore by amalgamation in sheds or open court-yards.
Beneficio de pella de plata—Applied when a portion of amalgam of silver is added to the mass under amalgamation.
Beneficio de la colpa—A method of amalgamation in which, instead of *magistral*, *colpa*, (colcothar) is used.
Beneficio per fuego—Reduction of ore by smelting.
Berilo—Beryl.
Bigorneta—A small anvil.
Bigornia—An anvil.
Bismut—Bismuth.
Blandura—Soft, crumbly ground.
Blenda—Blende, sulphuret of zinc.
Boca—Mouth, entrance, or pit of a mine; first opening made in the vein.
Boca mejora—A shaft or boca to com-

municate with the entrance of the vein to facilitate the workings.

Bochorno—Vapor, or foul air; want of ventilation; suffocating heat.

Boletas—Tickets of sales of ores; cheque tickets; account of charges, and produce of one amalgamation operation.

Bolsa—A purse; sometimes applied to a bunch of good ore of the supposed shape of a full purse or bag.

Bomba—A pump.

Bonanza—Prosperity; fine weather. A mine in *bonanza* is in a prosperous state; stopping costs; yielding profit.

Bordes—Border; ore left untouched by previous working in an old mine work.

Bordeta—A small pillar.

Borrasca—Adversity; foul weather. A mine in *borrasca* is in an unproductive state; does not stop costs.

Bota—A leather bag or sack, made of one or more skins, in which water is lifted in the mines.

Bota chica—Small leather sack.

Bota grande—Sack made of two or two and a half hides, used to extract water, and worked by horse whims.

Botalla de burro—Sack or bag, made of one neat hide, to extract water, and worked by a burro, or hand whim.

Botillo de lomo—Small sack made of one-third of a hide, to carry water out of small sinkings on men's backs.

Boveda—Vault, arch.

Braceage—Brassage in coinage.

Broculas—Drills.

Bronce—Brass, gun-metal, iron pyrites.

Buytron—Furnace for smelting ores.

Burilada—A chip taken from a mass of silver, to try whether it be standard.

Burro—A hand-whim, a windlass.

Busca—(Search.) The right to employ *buscones*, frequently claimed by the administration and persons employed in mines.

Buscones—Tributers, or miners who work on part proceeds; also, those who search for ores in a metalliferous district generally, or in a mine for such ores as have been neglected, and left behind by others.

C.

Caballerangos—Horsekeepers.

Caballitos—Men who carry the mining captains or others.

Caballo—A horse; a mass of the sterile mountain rock immersed in the lode.

Caballo de tepatate—A mass of barren rock interposed in a vein.

Cadena—A chain.

Cajon—In Peru and Chile two montons of 32 quintals each—64 quintals.

Cajon de granza—The pit to receive the crushed ore.

Cajonero—A lander, or one who receives the bota or manta at the shaft's mouth.

Cal—Lime.

Calde estaño—Calx of tin.

Calderas—Boilers.

Calenta dura—The first heating of the furnace, or putting the furnace into blast.

Calicada—Shode pit.

Caliche—Calcareous matters; ores of a calcareous character.

Calicheros—Lime burners.

Calientes—Warm ores (in amalgamation) containing sulphuret of iron and copper, and no calcareous matter.

Caliza, or *Cal en piedra*—Limestone.

Cal viva—Quicklime.

Campistas—Tributers.

Campo—A pitch, a working in possession of buscones, or allotted to the proprietors or others.

Canal—A spout, a canal.

Candallero—A socket deeper than the chumacera, and used for the same purpose.

Canella—Used to convey the fire to the charge for blasting.

Cañon—In a mine, a level, or horizontal gallery.

Caños—Pipes, tubes.

Cantera—A quarry.

Cantero—Stone mason, quarryman; a pitcher.

Caperossa—The person in charge of the sheds under which the tahonas are worked.

Capellinas—Large iron or silver bells, under which the quicksilver is separated from the silver in the amalgam by distillation.

Capellina—The iron bell under which the silver amalgam is distilled down. See *Piña.*

Carbon—Coal, charcoal.

Carbon de leña—Charcoal.

Carbon de piedra or *Carbon de tierra*—Pit coal.

Carboneros—Carriers, makers, and sellers of charcoal.

Carcamo—The drain which carries off the earthy matter from the tinas when washing the amalgam.

Cardenillo—Verdigris.

Carena—An upright stanchion for supporting machinery.

Carga—380 lbs. Spanish; a mule load, varying in the different mining districts; a charge for blasting; a lode.

Carilleros—Ore-carriers.

Carpentero—A carpenter.

Carretilla de mano—A wheelbarrow.

Carrita—A wagon or cart.

Carritero—A wagoner.

Casa de moneda—The mint.

Cascajal—A gravel pit.

Cascajo—Gravel, rubbish.

Caso—A boiler used in hot amalgamation.

Castillo—The frame of the stamping machine.

Castina—Fluor or flux.

Cata—A mine of no great depth; a pit made in quest of the vein; a mine denounced for trial.

Catear—To search for new mines.

Caxo—Money chest, treasury.

Caxo del tiro—Reservoir at the bottom of the shaft; sump of the pit; that part of the pit below the deepest level driven from it.

Caxo—A measure of ore containing many quintals, but varying in bulk at different places; at Potosi, 5000 lbs.; a handbarrow.

Caxoneros—Landers at the mouth of a shaft.

Cebar—To feed; to supply a furnace with materials for smelting; to add quicksilver to a mass of ores under amalgamation.

Cebo—Priming, as of gunpowder; a feed, as of oats for horse; second addition of quicksilver to the torta.

Cedazos—Sieves.

Centrada—The bottom of the refining or cupelling furnaces, made of very fine earth and vegetable ashes which have been lixiviated; materials of which a cupelling test is made; the test itself.

Cendradilla ó galeme—Cupelo.

Cerro—A mountain.

Cestas—Baskets.

Chapas—Iron blocks on which the stamps fall.

Charquedores—Cart-fillers.

Charqueo—Filling the baskets by hand.

Chihicles—Crystallized, calcareous, or other spar.

Chicuites—Baskets.

Chiflon—*Trabajar chiflon*—A phrase, meaning the work making way, both in length and depth.

Chino—Iron or copper pyrites.

Chumacerá—An iron socket for the socket of shafts.

Cielo—Roof of a work, an upward working. *Trabajar de cielo*—working at the roof or top of the vein.

Cilindro—A cylinder.

Cinabrio—Cinnabar, sulphuret of mercury.

Claro—An open space on the lode from which ore has been taken.

Clavos—Masses of native metal; bunches or masses of ore; nails.
Cobalto—Cobalt.
Cobre—Copper.
Cobre en roseta—Rose copper.
Cobriso—An inferior kind of copper ore.
Cohete—A cartridge for blasting.
Colero—Assistant of the underground captain in charge of the peonada, or account of the daily labor.
Collado—A hill.
Colorados—Ores colored with red oxyd of iron. See *Pacos*.
Comer los pilares—To take away the pillars of the lode which had been left during the previous working of the mine to support the roof, and consequently to abandon the mine.
Comerse los pilares—The same as *Comer los pilares*—figuratively, to abandon a mine.
Comillo—A reverberatory furnace.
Compromiso—A private engagement or undertaking also a joint-stock undertaking.
Conducta—A convoy or caravan, conveying the precious metals or coin overland.
Contra cielo—A rise, or working upwards.
Contra mina—A work of communication between two mines; also an adit.
Contra tiro—Auxiliary pit contiguous to a main pit or shaft, to serve as a footway, or for ventilation.
Convenio—A legal agreement.
Copola—A cupelling furnace.
Copos or *Pazillas*—In amalgamation, little globules into which the quicksilver forms when the process is too quick.
Coral—Copper.
Cortar pilar—To complete the pillar, and make a cross passage, and to form a landing place.
Cortar las sogas—To abandon a mine; *lit.* to cut the ropes.
Costadores—Wood-cutters.
Costal—Sack for ore, made of the *pita* or thread of the aloe; a rammer or beetle.
Cras—An iron cage or frame used in smelting.
Creston—Out-cropping of a lode; the crest of a lode appearing above the surface of the ground.
Criadero—A spot which is metalliferous; a spot, a district, a mountain, a rock, where ores are supposed to grow.
Criba—Perforated leather, through which the stamped ore falls into a pit.
Crisolito—Crysolite.
Cristal—Crystal.
Cristal de roca—Rock crystal.
Crucero—A cross cut.
Cruces—The cross pieces of the arrastras or grinding mills.
Cruzada—Applied to lodes, that which is crossed by another.
Cubo—A leather or other bucket.
Cuchara—The scraper used for extracting the pulverized ore or rock in the hole made to receive the charge for blasting; a spoon.
Cuchara de cuerno—A small horn bowl in which the earth of the tortas is washed, in order to ascertain from time to time the progress of the amalgamation.
Cuele—The art or effect of cutting or driving a mine; work in any direction; the distance or space advanced in a mining work.
Cuerda—A rope.
Cueros—Skins, generally of oxen, cows, etc.
Cuerpo—The lode; also body, as *cuerpo de la veta*, body or main vein of the lode.
Cuerpo alto—Upper branch of the lode; upper story of a house or store.
Cuerpo bajo—lower branch of the lode; ground floor of a house or store.
Cuerpo medio—Middle branch of the

lode. These three terms are used where a lode is divided into three parallel branches, with the same underlay, as at Veta Grande, Guanaxuato.

Cuña—A wedge.

Curtir—The operation of adding lime to warm ores, or magistral to cold ores, in amalgamation.

D.

Dedo—The twelfth part of a palmo; four palmos making one vara; twelve dedos=eight pulgados, or Spanish inches.

Demasia—Space unappropriated between mines or otherwise.

Denuncio—*Denunciacion*—Denouncement; a formal application to the court of mines of the district to have a mine adjudged to the applicant, the workings of which have been abandoned or inefficiently carried on during the period fixed by the ordinance, or which has never been worked at all. A person has the power of denouncing a mine which has been unworked, or inefficiently worked, or depopulated for more than four months.

Deputacion de mineria—Mining tribunal.

Derrecho—Straight, right.

Devrumbe or *Derrumbamiento*—The falling-in of the works of a mine.

Desaguador—Water-pipe or conduit.

Desagüe—Drainage.

Desagües—Outlets of every description by which the water is got rid of.

Descargue—Drawing out the last contents of the furnace; blowing out the furnace.

Descostradores—Men employed in taking down any fragment which may remain after blasting.

Descubridora (*mina*)—The mine in which the vein has been discovered, and which is entitled by law to a double *pertenencia*, if in a district already worked, or a triple *pertenencia*, if in a new district.

Desfrutar—To enjoy, to have the benefit of a mine.

Desmonte—Clearing away the surface of the ground; removing by the pick, blasting, or otherwise, the mountain rock, or breaking down ores.

Despachadores—Men employed in filling the *mantas* with ores, etc.

Despacho—Plat; point of junction between a shaft and a level enlarged for receiving the ores, etc., to be sent up the shaft to the surface.

Despensa—A store-room for materials and tools; sometimes, also, for the quicksilver, and even for the bullion.

Despueble—Abandoning the mine, or omitting to keep the proper number of hands at work.

Destajo—Piece or contract work; tut work.

Destajero—Tut workman, or one who undertakes to work by contract.

Diamante—Diamond.

Dique—A dike.

E.

Echadero—A plain where the mules are loaded, the metal spread out, cleaned, and weighed.

Echado—Inclination or dip of the slope.

Ege—The axis of a wheel, axis of a carriage, etc.

Embolo—A piston.

Emborrascarce—Applied to a vein which has become barren, as the spot then in work.

Emparejar—To level or square; harness cattle, saddle them.

Empleo—The quantity of quicksilver mixed with the ore on any given occasion for effecting the amalgamation.

En bonanza—Yielding profitable returns.

Encampanar una mina—To cut off the workings of a mine on the underlay by working on the lowest works from a neighboring mine.

Encapillar—To form a chamber, or an enlargement of a working, preparatory to driving another work from it.

En frutos—In produce; producing ores.

Ensalmorar—The first process in amalgamation, the act of mixing the saltierra with the lama.

Ensayador—Assayer.

Ensaye—Assay.

Ensayo—A trial.

Escaleras—Ladders made of poles of timber with notches cut in them, or otherwise.

Esmanil—Blende.

Esmeralda—Emerald.

Esmeril—Emery.

Espato—Spar.

Espato calizo—Calcareous Spar.

Espato fluor—Fluor Spar.

Espato pesado—Heavy spar; sulphate of lime.

Espejuelo—Mica.

Espeque—The cross level of the norea or tahona to which the mules are harnessed; a lever.

Estaca fixa—The post driven into the ground, from which the pertenencia is originally measured. *Estaca* means a stake.

Estado—A statement or account.

Estaño—Tin.

Estanique—Pond; dam of water.

Estoraque—Brown blende; sulphuretted zinc.

F.

Faenas—Work done by common laborers, such as dead work, removing rubbish, etc.

Faenero—Rubbish carrier.

Fanega—A dry measure containing 12 celamines, or 1.599 of an English bushel.

Fanegado—An extent of land; 90⅓ fanegados are equal to 100 English acres.

Feldspato—Feldspar.

Fierros—Stein, vulgarly Regulus, from the smelting furnace.

Flete—Freight.

Fluorspato—Fluor Spar.

Fluxo—Flux.

Fondon—A furnace for smelting ores.

Fosforo—Phosphorus.

Fosiles—Fossils.

Fragua—A forge.

Frios—In amalgamation, cold ores, those containing calcareous matter, and therefore requiring a larger quantity of sulphuric acid from the magistral.

Frente—An end; a forehead; an extremity of an adit or other level.

Frijolillo—A breccia.

Frijoles—Frijoles, French beans, the common food of the country.

Frutos—Produce; ore.

Fuellos—Bellows.

Fundidor—A founder; a smelter.

Fundicion—Smelting; also the smelting house.

G.

Galena—Galena, sulphuret of lead.

Galera—A large shed; a mill-house, or grinding-mill; a large building, on the floor of which the treading in of the quicksilver in amalgamation takes place.

Galleria—A gallery; a level.

Gallos—Small particles of silver, which appear in the shape of spherules on the surface of certain ores after they have been strongly heated; more generally, however, applied to the spurting out of the silver from the assay button on cooling; also, fine specimens of native silver, or other rich ores.

Gamela—A large wooden bowl.

Galpeador—A miner who works with the mallet or hammer in blasting.

Granada—Garnet.

Granito—Granite.

Grano—A grain.
Granos de oro—Grains of gold.
Granza—Coarse particles of ore after grinding which require to be ground again; brayed ore.
Granzas—Poor ores.
Grasas—Slag from the smelting furnace.
Greda—Chalk.
Greña—Ores in the rough state, not cleaned.
Greta—Litharge, fuller's earth.
Guarda—A rib of different substance from the rock or lode, which, generally, is upon the sides of the vein; called in Cornwall cupels of a lode, or backs.
Guarda raya—Marks or limits of the boring, or measurement of the work done in a mine; limit or boundary line.
Gueldra—The large cross-beam in which the upper spindles of the shafts of machinery traverse.
Guia—A mark directing to the richest part of the vein.
Guijo—The iron spindle of the shaft of machinery.
Guixa—Quartz.

H.

Hallilitador—He who supplies money for working a mine.
Hachas—Hatchets, axes.
Hacienda—Farm; estate; establishment for reducing ores.
Hacienda de beneficio—Establishment for reducing ores.
Hacienda de fundicion—Establishment for smelting ores.
Haciendero—The superintendent of the hacienda.
Hechado—Dip of the lode.
Herra mienta—Tools; taken figuratively, it implies a borer and hammerman.
Hierro—Iron.
Hierro colado—Cast iron.
Hierro labrado—Wrought iron.
Hilo—A small vein or thread of ore in a lode.
Hilo de la veta—Line or direction of the vein.
Hilos altos—Threads or small veins of ore falling into or proceeding from the upper or hanging wall of a lode.
Hilos bajos—Threads or small veins of ore proceeding from or falling into the lower wall of a lode.
Hoja de lata—Tin plate.
Hoja de laton—Sheet brass.
Hoja de libro—Finely laminated clay; slate; talc; *lit.* leaf of a book.
Horno—A furnace.
Horno de fundicion—A smelting furnace.
Horno de magistral—Roasting stove for copper pyrites.
Huaco—A hollow.
Hueco—A hollow place.
Huembas—Small rough beams of buildings.
Hundido—Sunk in; workings which have fallen in.

I.

Incorporar—In amalgamation, to add the first charge of quicksilver. The term *cebar* is applied to adding the subsequent charges; it also means the act of mixing in thoroughly the whole of the quicksilver with torta of wet ore.
Ingenios—Engines.
Instrumentos—Instruments, tools.
Intendente—Intendant.
Interventor—Inspector, representing the interests of the proprietors by whom appointed, or of the *aviador*.
Iridio—Iridium.

J.

Jantilla—A double-handled ladle, into which the melted silver falls from the cras.
Jaspe—Jasper.

Jornaleros—Day laborers.
Jorongo—A small basket; also a blanket.

L.

Labor—A work from which ores are extracted in general, all work of the mine, and especially the front work.
Labores de hacienda—All workings in a mine not let to tributers.
Ladrillera—An iron or stone mould, into which the melted silver is poured in order to form the barra.
Ladrillas—Bricks.
Lama—Slime or schlem from the amalgamator.
Lamero—The lama when merely thickened by admixture with saltierra.
Lameros—Lama pits.
Lancha—A sort of hard freestone.
Lapiz—Black lead.
Lapiz encarmado—Red chalk.
Laton—Brass.
Lavador—A man employed in washing the ore after amalgamation, or rather in cleansing the amalgam.
Lavaderos—Gold washings; washing vats or tubs for separating the amalgam from the slime.
Lazadores—Men who procure people to work in the mine; also, men employed to catch cattle.
Leña—Firewood.
Leñadores—The workmen employed to carry and serve the wood to the smelting furnace; also, the wood-cutters, collectors and sellers of firewood.
Lenter nillas—Large vertical wheels of the stamping apparatus.
Ley—Standard of the metals; contents in pure silver.
Ley de oro—Quantity of gold contained in the silver.
Ley de plata—Quantity of silver contained in the ore.
Libramiento—Warrant for payment for bars of gold or silver delivered at the mint, or order for funds.
Libranza—A bill of exchange.
Ligra—Flux.
Lima dura—An appearance put on by the quicksilver in certain stages of the process of amalgamation, which is noticed at the edges of the amalgam washed in the bowl for making a *tentadura*, or trial.
Limpia—Clearing out of rubbish and ruins from the old work in a mine.
Lis—A particular state of the amalgam, observed by means of the *tentaduras* or trials in the bowl.
Llano—A plain, flat ground.
Llevada—Carriage, transport.
Llevador—Carrier, conductor.
Lode—Mud.
Losa—A flat stone.
Lumbrera—An air shaft; an adit shaft.

M.

Macero—He who has the charge and direction of crushing and grinding the ore in the tahonas previous to amalgamation.
Macizo—A solid, untouched part of the vein.
Madera—Timber.
Maestro herrero—Master blacksmith.
Maestro carpentero—Master carpenter.
Magistral—Copper pyrites used in amalgamation.
Magistral—Roasted copper ore for mixing in the amalgam heap.
Malacate—A horse whim.
Malacatero—A whim driver.
Maiz—Indian corn, the principal food used at the mines.
Malacate doble, or *Malacate sencillo*—The former whim has bags made of two ox-hides holding 1,250 lbs. water; the latter, one hide, and holding half the quantity.
Manantial—A spring of water.
Mandadero—Errand boy.
Mandon—Master miner, or overseer.
Manganesa—Manganese.
Manta—A blanket, or horse cloth, used to contain ores or tools to be brought

up by the malacates, now replaced generally by sacks made of the fibers of the agave, or ox-hides.

Mantear—To raise ore from the shafts in bags or mantas.

Manto—A bed or circumscribed stratum.

Maquila—Rate paid to the proprietor of a mill, or reduction work, for reducing ore on another party's account; applied chiefly to reduction by amalgamation.

Maquilero—One who dresses ores for hire.

Maquina—A machine.

Maquinas de vapor—Steam engines.

Marco—8 oz., or lb. Spanish, equal to 3,552 grains English.

Marco de oro—8 ounces of gold.

Marco de plata—8 ounces of silver. The marco de oro or marco de plata may be standard, or otherwise.

Marmol—Marble.

Maroma—A rope to pull or draw by, as a hawser.

Marquesitas—Mundic; iron pyrites.

Martillo—A hammer.

Martriquila—A register for miners, etc.

Maza—Stamp head for pounding the ores.

Mecha—A match or fuse.

Medida—A measure.

Mejora—Improvement.

Mejora de boća—A term used when an improvement or alteration is made in the entrance to a mine.

Memoria—Weekly account of disbursements and mine expenses.

Mercurio—Quicksilver.

Merma—Loss of quicksilver in amalgamation, or of lead in smelting.

Meson—An inn mostly appropriated to muleteers.

Metal—Metal, ore.

Metal de ayuda—Metal or ores added in smelting, to assist in the reduction of the silver ores, lead, or galena, for example.

Metal pepena—Selected and picked gold and silver ores.

Metales communes—Common ores.

Metales de fundicion—Ores for smelting.

Metales plomosas—Ores impregnated with lead.

Metales porosos—Porous ores.

Metapiles—Grindstones used in the tahonas; also, pigs of copper used in hot amalgamation.

Mineral—Ore; mineral; recently applied to a mining district, formerly and still called Real de Minas.

Mineria, deputacion de—A tribunal cognizant of mining matters, elected in most cases by the mine-owners of the district.

Minero—A miner; an underground agent.

Minio—Red lead.

Modelos—Models.

Mojon—A land-mark used to mark the limits of pertenencias.

Molibdena—Molybdena.

Molienda—The act of grinding or pounding the ores; sometimes used to designate the ores ground. "*La Molienda.*"

Moliente—Shaft of tina.

Molino or *Mortero*—Stamping mill.

Molongues—Crystallization of silver ores very rich.

Montaña—A mountain.

Monteros—Stampers.

Montes—Woods.

Monton—A heap of ore; a batch under the process of amalgamation, varying in different mining districts. At Cotorce, 36 quintals; at Guanaxuato, 35 quintals; at Real del Monte, Pachuca, Sultepec and Tasca, 30 quintals; Zacatecas and Somburete, 20; Presnillo, 18; Bolanos, 15; and at Valenciana, 32.

Monton—Small heaps of ore mud, for amalgamation.

Mozo—A man-servant.

Muestras—Samples.
Mufla—A tuyere.

N.

Natas o Escorias—Slags.
Natron—Native carbonate of soda.
Negociacion—Business undertaking, as a mine, or set of mines, etc.
Nicolo—Nickel.
Nitro—Nitre.
Nivel—Level.
Nivels de agua—Water levels.
Noria—An endless chain, with buckets attached, revolving round a wheel; it is used underground for drawing water out of the pozos or sinks, which are carried down to a greater depth than the principal shaft; also, a common superficial machine for raising water.

O.

Obras—Workings.
Ocre—Ochre.
Ocre rojo, or ocre colorado—Red ochre.
Official de carpentero—Journeyman carpenter.
Official de albañil—Journeyman bricklayer.
Ojo—Bunch or small spot of ore in a lode.
Ojo de polvillo—Spots of rich ore.
Ojo de vibora—Black sulphuret of zinc.
Onique—Onyx.
Opalo—Opal.
Operanos—Workmen.
Ordenanzas de mineria—Code of mining laws.
Oro—Gold.
Oro de copela—Fine gold.
Oro empolvedo—Gold dust.
Orphimento, or *Oripomonte*—Orpiment.

P.

Pacos—Earthy ores, consisting of oxyd of iron mixed with various ores of silver; when of a red color, they are frequently called *colcorados*; they are generally found near the surface.
Paja—Straw.
Pala—A wooden shovel.
Paladion—Palladium.
Palanca—A lever; a pole on which a weight is supported by two men.
Palanza de hierro—Crowbar.
Palmo—Quarter of a vara, or Spanish yard.
Pannio—The ground or country through which the lode runs; also, the matrix.
Panizo—Hornstone.
Parado—A relief or change of men, mules, or horses.
Parcinera—A partner in the mines.
Parihuela—A letter.
Partido—Division of ores between the owners and buscones.
Pasta—Uncoined silver.
Patio—A yard, court; floor of a court on which the ores pass through the process of amalgamation.
Patio—Amalgamation floor.
Pegador—Man who sets fire to the matches for blasting.
Pella—The silver mixed with quicksilver, when all the latter metal has been forcibly pressed out, except the portion which can only be separated by distillation.
Peltre—Pewter.
Peones—Native laborers or assistants; day laborers.
Pepena—Picked ore of the best quality; rich ore.
Pepenado—Cleaned ore.
Pepenadores—Cobbers, cleaners, and classers of the ores.
Pepitas—Small grains of native silver or gold.
Peritos—Intelligent or practical persons selected as arbitrators to decide scientific or practical questions or disputes, or to determine the underlay of veins prior to fixing the limits of the *pertenencias*.

Perla magarita—Pearl.

Pertenencia—Extent of 200 varas upon the course of a lode to which a title is acquired by denunciation; the breadth varies, according to the underlay of the vein, from 112½ varas to 200 varas.

Peso—A dollar; any weight.

Petlanques—Crystallization of silver ores; also, silver ores which are very conspicuous in the matrix; for example, *petlanque colorado* is the red antimonial silver, whether crystallized or otherwise.

Pez—Pitch.

Pico—A miner's pick.

Piedra—Stone.

Piedra de toque—Touchstone.

Piedra cornea—Hornstone.

Piedra iman—Loadstone.

Piedra podrida—Rotten stone.

Piedra pomez—Pumice stone.

Piedras de mano—Good pieces of ore, sometimes carried up by hand, and often assigned to pious purposes.

Piedras preciosas—Precious stones.

Pilares, or *pilarejas*—The pillars of a mine.

Pileta—A trough; the hollow basin before the smelting furnace into which the metal flows; tank or small reservoir underground to collect the water of infiltration.

Piña—The cake of silver left after the quicksilver has been distilled off.

Piña—Piles of amalgam for distilling down the mercury.

Pinta—The appearance, whether favorable or unfavorable, of a fragment detached from the lode; the mark of particular metals, by which their value is recognized according to their appearance to the eye.

Pintar—To exhibit pintas, or indications of ores.

Pirites—Sulphuret of iron.

Piso—The bottom or floor of a work.

Pison—A rammer.

Pita—Thread made of the fiber of the agave or maguey.

Pizarra—Slate.

Plan—A bottom working, or working driven from the bottom of a level, adit, etc.

Plancha—Pigs, as *plancha de plomo*, pigs of lead.

Plata—Silver.

Plata de ley—Standard silver.

Plata piña—Silver after distilling off the mercury.

Plata piña—The porous silver cakes left after distilling down the mercury.

Plata parda azula y verda—Muriate of silver of different colors.

Platina—Platinum.

Pleito—A law suit.

Plomo—Lead.

Plomosos—Applied to ores containing lead.

Poblar—To set on workmen in any mine.

Polvillones—Rich ores.

Polvillos—Applied to ores, tender, rich.

Polvillos buenos—Good ores of the kind.

Polvo—Dust.

Polvoro—Gunpowder.

Polvorilla—Black silver; disseminated sulphuret of silver.

Porfido—Porphyry.

Potasa—Potash.

Pozo—A sink on the inclination of the vein; a pit; a well.

Presa—A dam.

Protocola—Minutes.

Pueble—Actual labor performed in a mine, with the number of workmen at least prescribed by the mining laws.

Puertas—Very strong rock, which conceals the vein, and which requires blasting ere the vein is discovered; also, doors.

Pulgada—An inch.

Q.

Quadrado—A square.
Quajado—Dull lead ore.
Quarzo—Quartz.
Quebrada—A ravine.
Quebradores—Cobbers or breakers of the ore; men who break up the ores on the surface.
Quemadero—Burning house or place.
Quemazon—The barren, scorched appearance of the crest of a metalliferous lode protruding from the surface of the mountain.
Quilate—Synonymous with carat; for example, gold of 22 quilates contains 22-24 parts of pure gold, just as the English standard gold of 22-24 parts of pure gold. The quilate is divided into four granos Spanish.
Quintal—4 arrobas, or 100 lbs. Spanish, equal to 101 45-100 lbs. English.
Quita pepena—A man who stands at the mouth of a shaft, to see that none of the metal is stolen.

R.

Ramo—A branch from the main vein.
Rancho—A detached farm-house and ground; the house is often nothing more than a mere hut.
Raya—Weekly account of the expenses.
Rayador—Clerk who keeps account of the workmen's time, the stores received, etc.
Real—⅛ of a dollar; a mining district.
Real de minas—The term generally applied to a mining district, although mineral de minas is also now used.
Reata—A rope about as thick as a finger, or larger, used as lashings to *cargas.*
Reatilla—A single-twisted smaller rope.
Rebaje—A working down of high ground.
Reboltura—A mixture of the ground ore with the usual reagents or fluxes.
Rebosadero—Crest of a vein.
Recina—Rosin.
Recuesto—Inclination of a vein.
Regador—One who has a right to a certain share of water for irrigation.
Registrar—To get an entry made by the proper officer of a party taking possession of a new mine.
Registro—An entry, as above described.
Reliz—The wall of the vein.
Remolino—Mass or bunch of ore.
Rendvise—In amalgamation, said of a torta when the amalgamation is complete.
Repasar—The mixing together of the ore, quicksilver, and other ingredients in the wet state, in order to extract the metal; to work the quicksilver into the tortas of ore under amalgamation, done by treading them with mules or men.
Repasador—Laborer who treads the quicksilver into the ores under amalgamation.
Rescatador—Ore purchaser.
Rescate—Public sale of ore.
Retortas—Retorts.
Rio—A river.
Riscos—Crystals.
Rosicler—Ruby silver; red antimonial ore.
Rubi—Ruby.
Ruedas—Wheels.
Rumbo—Point of the magnetic compass.
Rumbo de la veta—Line or direction of the vein.

S.

Saca—The ore obtained from a mine in a given space of time.
Sacabocados—Punches.
Sacabuches—Hand-pumps.
Sal—Salt.
Sala—The principal room of a hacienda, or any other building.
Salitre—Saltpetre.
Sal mineral—Mineral salt for amalgamation.
Sal tierra—Impure or earthy salt.

Sal tierra—Salt earth, (containing 12 or 13 per cent salt) for mixing in the amalgamation heap. See *Torta.*

Salineros—Applied to ores requiring much salt in amalgamation.

Salones—Saloons, large halls, hollows, or cavities in a lode.

Sangria—Letting off water by piercing the substance which dammed it in.

Sanguinaria—Blood-stone.

Sebo—Tallow or suet used for machinery, etc.

Serape—Blanket, the usual dress of miners.

Sierras—Saws; chains of mountains.

Silla—A saddle; a leather which passes over men's shoulders to protect them in carrying the ore.

Sitio de labor—Land measure, 5,000 varas in diameter.

Sobrante—Surplus, profit, residue.

Socabon—An adit or water level driven from the earth's surface, either on, the course of a lode, or to intersect it.

Soga—The rope to which the bota is attached.

Soguilla—The rope for the botillas de burro, botas chicas, or mantas.

Sòlda dura—Solda.

Soliman—Corrosive sublimate.

Sombra—Shade; gray tinge of certain ores or matrices of ores.

Soplete—A blow-pipe.

Soto minero—Sub-miner.

Sucino—Amber.

T.

Tablas—Planks.

Tahoma—A mill of small horizontal stones.

Tajadera—Wedge to break the tinas.

Tajamanil—Shingle for roofing.

Tajo—A cut.

Tajo abierto—An open cut.

Talco—Talc.

Talegra—Bags of dollars, containing 1,000.

Tanda—Task; compulsory labor; days appointed for working the mine; duration of the period in which a regador is entitled to use running water for irrigation.

Tanque—Tank, or artificial pond.

Tantalio—Tantalum.

Tapa ajos—Bandage for the eyes; used to cover the eyes of mules when treading the ore in the patio, or elsewhere.

Tarea—A task; also a certain quantity of wood for fuel; the quantity cut upon one task.

Tejo—Cake of metal.

Tellurio—Tellurium.

Tenates—Sacks made of pita, or hide, for raising the ore on men's shoulders; large leather, or coarse linen bags, in which the tenateros remove the ores and rubbish.

Tenatero—Ore-carrier from the workings to the surface, or to the despacho only.

Tentadura—An assay, or trial.

Tepetate—Rubbish remaining after cleaning the ores; also applied to all the earth of the mine which contains no metal; barren rock through which the ore runs.

Terquisite—Native carbonate of soda.

Terreros—Heaps of attle and rubbish from the mine.

Tersoreria—Treasury.

Testera—A dyke interrupting the course of a lode.

Texear bienen horno—When the litharge is thrown out from the furnace.

Tienda de raya—A shop at which the miners obtain weekly credit.

Tierra pesada—Barytes.

Tierras—Earths; applied to ores, earthy, poor.

Tierras apolvillados—Ores a degree inferior to azogues apolvillados.

Tierras communes—Common earthy ores.

Tierras de morlero—Poor stamped ores.

Tina—A vat, or jar.

Tiro—A shaft.
Tiro de mulas—A set of mules.
Tiro general—The principal shaft.
Titanio—Titanium.
Tornero—A wooden vat.
Torta—A certain quantity of ore made under amalgamation, forming one heap, which, being of a flat shape, is called a torta, or cake.
Torta—The great flat heap of silver ore for amalgamation.
Torta rendita—Amalgam ready to undergo the washing operation.
Trapiche—Grinding mill.
Trementina—Turpentine.
Triangulos—The cogs of the stamps.
Tribunal de mineria—Mining tribunal.
Trompeto—A small malacate.
Tungstena—Tungsten.
Turba—Turf, peat.
Turbit mineral—Yellow oxyd, a sulphuret of mercury.

U.

Uranio—Uranium.

V.

Vapor—Steam; foul air in a mine.
Vara—A Spanish yard, equal to 33 inches English, nearly—109 30-100 varas equaling 100 English yards.
Velador—A watchman who takes care of the mine day and night; an under miner.
Velas—Candles.
Vena, or *Veta*—A vein, or lode; it is called *manta* (a cloak) when it is a bed; *clavada* (upright) when it is vertical, or nearly so; *echada* (inclined) when it has a certain dip, or inclination; *obliqua* (oblique) when it goes in that direction through the mountain; *serpenteada*, when in a serpentine direction; *socia* (companion) when it is joined by another; *rama* (branch) when it proceeds from the main vein.
Verde tierra—Verditer.
Vermellon—Vermillion.
Veta madre—The mother, or principal vein.
Vidrio—Glass.
Vigas—Beams, split or sawed timber.
Vitriolo—Vitriol.
Vitriolo azul—Blue vitriol.
Vitriolo blanco—White vitriol.
Vitriolo verde—Green vitriol, or copperas.
Voladiras—Grinding stones at the arrastras.
Vueita—The glow of the silver, in cupellation, when the last film of oxyd of lead suddenly separates and disappears.

X.

Xabon—Soap; a peculiar description of ores.
Xabones buenos—Good ores of the above description.
Xacal—A hut in which ores and tools are kept.

Y.

Yesca—Tinder, or touchwood.
Yeso—Sulphuret of lime.
Yungue—Anvil.

Z.

Zacate—Maize, straw, or grass, given to the mules.
Zanca—A ditch.
Zurron—A sack made of leather; cochineal is packed in zurrones.

Bibliography

Comment on Sources

In general, the bibliography of the mining West can be divided into four major categories. Two of them are reputable and complementary. The third is a curate's egg, parts of which are quite good. The fourth would be better ignored. The professional technical studies and reports de-emphasize the humanities but are indispensable materials by educated men who were on the spot. With but few exceptions they are completely reliable. Contemporary popular accounts and the scholarly secondary histories play down technology but are correspondingly strong in humanistics. What cannot be found in the one range of works can usually be discovered in the other, and the careful historian consults both. Mining history has fortunately been spared the attentions of fiction writers and special pleaders, as well as the spurious memoirists, such as those who weave a tangled web about the open-range cattle industry or the last stand of the American Indians.

It is hard to make general assertions about the institutional histories of mining corporations. Many contain information not easily gained elsewhere, but it is evident that their sponsors were not moved by the same consideration as those that motivate professional historians. Contemporary newspaper accounts are reliable in direct proportion to the knowledge and veracity of the reporters, two highly variable factors. Journalists and such promotional writers as J. Ross Browne were often quite frank about developments, if less than candid about future possibilities. Each account therefore represents an individual case which must be assessed upon its own merits and in the light of historical hindsight.

The pointed absence of certain works from the bibliography that follows, together with some hints about the nature of those that are included, may enable the student to economize on time by ignoring them. In this ragged literary regiment marches first the "lost mines and buried treasure" genre, the better examples of which may be of interest to folklorists or of use as guide to superstitions but are otherwise worthless. Next comes modern journalistic popularization, usually based upon the cursory reading of a few major sources. Such works are likely to be deceptive in interpretation, breathless in manner, and filled with errors. The "ghost town" album or "picture book" may have interesting illustrations, but its information may be inaccurate. To give one illustration, a pamphlet sold in Virginia City states that the Spaniards worked the Comstock Lode long before it was located in 1859, as instanced by the presence of Spanish mining methods and tools. This statement, of course, ignores the readily discoverable fact that the Mexican Mine was so worked but was contemporary with its fellows of the flush times.

Even supposedly reputable sources must be scrutinized with care so as to be certain that the occasional incredible assertion was the result of misunderstanding and is not evidence of forgery. I am grieved to report that Thomas Arthur Rickard was not immune to carelessness with historical fact, probably because he relied upon his memory instead of upon documents.

General Works

Bancroft, Hubert Howe. *History of Arizona and New Mexico, 1530–1888.* Facsimile ed. Albuquerque, 1962.

This work and those following are part of the thirty-nine volumes of history, chiefly documentary, of the western states, the product of Bancroft's great history factory. Students of a given region are referred to the volume or series of volumes covering the area. Although it is a lode, a few of the documents should be approached with caution.

———. *History of California.* 7 vols. San Francisco, 1888.

———. *History of Nevada, Colorado, and Wyoming.* San Francisco, 1890.

Fisher, Vardis, and Opal L. Holmes. *Gold Rushes and Mining Camps of the Early American West.* Caldwell, Idaho, 1968.

An album of photographs and illustrations accompanied by a cursory text. The bibliography is extensive but not carefully selected.

Greever, William S. *The Bonanza West: The Story of Western Mining Rushes, 1848–1900.* Norman, 1963.

Unquestionably the leading scholarly survey of the American mining frontier, emphasizing social history and economic developments.

Paul, Rodman W. *Mining Frontiers of the Far West, 1848–1880.* New York, 1963.

Like Greever's book, but with slightly less intensive treatment, this general survey possesses an excellent critical bibliography.

Rickard, Thomas A. *A History of American Mining.* New York, 1932.

Rickard was the leader in mining engineering in the period 1890–1920, a voluminous writer, and the editor of the three professional mining journals. Despite its unevenness this book is a fundamental study, based in part upon technical articles, upon historical research, and upon Rickard's memory. It contains a few small errors of historical fact against which the student should be on guard.

———. *Man and Metals: A History of Mining in Relation to the Development of Civilization.* 2 vols. New York and London, 1932.

This study, Rickard's chief contribution to history, was supervised by several leading professional historians, among them Sir Flinders Petrie. Although its interpretations are now outmoded, it is a valuable compilation and is in such demand that copies are virtually unobtainable.

———. "Rich Ore and Its Moral Effects," *Mining and Scientific Press,* Vol. XCVI (June 6, 1908), 774–75.

———. *The Romance of Mining.* New York and London, 1945.

A product of Rickard's old age, this book is superficial and scarcely rewarding except as a source from which to placer a few nuggets of fact omitted or overlooked in his earlier works.

———. "Salting: The Nefarious 'Art' and its Age-old Results as Practiced on the Mining Fraternity," *Engineering and Mining Journal,* Vol. CXLII (1941), 42–52.

Sloane, Howard N., and Lucille Sloane. *A Pictoral History of American Mining from Pre-Columbian Times to the Present Day.* New York, 1970.

An album of illustrations which, like the Fisher and Holmes work, is broad but not remarkably authoritative. Photographs seldom lie, however, and in its illustrations lies its chief appeal.

Sullivan, Mark. *Our Times.* 5 vols. New York and London, 1934.

Todd, Arthur C. *The Cornish Miner in America: The Story of Cousin Jacks and Their Jennies and Their Contribution to the Mining History of the American West, 1830–1914.* Truro, Cornwall, and Glendale, 1967.

Contains many sections and passages of great value, but too often is merely a compilation of the histories of Cornish families in the West. Recommended with reservations.

U.S. Bureau of the Census. *Abstract of the Fourteenth Census of the United States.* Washington, D.C., 1923.

———. *Thirteenth Census of the United States: Abstract, . . . with Supplement for California.* Washington, D.C., 1913.

Geological and Metallurgical References

Aitchison, Leslie. *A History of Metals.* 2 vols. New York, 1960.

An extremely valuable and well-documented history of metallurgy from ancient times. It serves as a fine companion piece to Rickard's *Man and Metals,* and is more soundly interpretative.

Bain, H. Foster, ed. *Types of Ore Deposits.* San Francisco, 1911.

Boldt, Joseph R., Jr. *The Winning of Nickel*. New York, 1967.

Although this excellent volume concerns a modern industry using ultra-modern methods, it has valuable sections on the evolution of drift mining from empirical to scientific methods.

Bray, J. L. *Nonferrous Production Metallurgy*. New York and London, 1947.

Bullard, Fred M. *Volcanoes in History, in Theory, in Eruption*. Austin, 1961.

Butler, G. Montague. *Some Facts About Ore Deposits*. Arizona Bureau of Mines *Bulletin* 139. Tucson, 1935.

Although this monograph was intended for the enlightenment of the small mine operator or would-be investor, its discussion of the rise of false theories of mineralization makes it essential to the student of the mining frontier.

Dennis, W. H. *A Hundred Years of Metallurgy*. Chicago, 1963.

A useful sourcebook for metallurgical inventions, their inventors, and such methods as the refining of matte at Swansea, Wales.

Egleston, Thomas. *The Metallurgy of Silver, Gold, and Mercury in the United States*. 2 vols. New York, 1887–90.

Egleston's scholarly work is intended as a handbook for the engineer but also contains many valuable historical references.

[Ehrenberg, Herman]. "The Reduction of Silver from Its Ores," *Weekly Arizonian* (Tubac), Vol. I, No. 22 (July 21, 1859).

Emmons, William H. *Gold Deposits of the World, with a Section on Prospecting*. New York and London, 1937.

The chief authoritative survey of the world's gold mines and placers, essential for understanding the prospecting and development of any gold-mining district.

Gaudin, A. M. *Principles of Mineral Dressing*. New York and London, 1939.

Hoffman, Arnold. *Free Gold: The Story of Canadian Mining*. New York, 1947.

Jagger, Thomas A. *My Experiments with Volcanoes*. Honolulu, 1956.

Lavender, David. *The Story of Cyprus Mines Corporation*. San Merino, Calif., 1962.

Liddell, D. M. *Handbook of Nonferrous Metallurgy*. 2 vols. New York and London, 1945.

McKinstry, Hugh E. *Mining Geology*. Englewood Cliffs, N.J., 1948.

Riley, Charles M. *Our Mineral Resources*. New York, 1959.

Shelton, John S. *Geology Illustrated*. San Francisco and London, 1966.

A magnificent, authoritative volume which describes the major landforms and illustrates them with beautiful aerial photographs and clear line drawings. A work of art as well as of science.

Tarr, Ralph S. *Economic Geology of the United States*. New York, 1904.

Taxay, Don. *The United States Mint and Coinage*. New York, 1966.

Mining Textbooks, Histories, and Memoirs

Barton, Denys B. *Cornish Beam Engines*. Marazion and Penzance, Cornwall, 1966.

Bennett, Russell H. *Quest for Ore*. Minneapolis, 1963.

Burgess, T. A. "Explosion in Compressed-Air Main," *Mining and Scientific Press*. Vol. XCVII (August 22, 1908), 253.

Chapman, Thomas G. "Treating Gold Ores," University of Arizona *Bulletin* 138, 1935.

Crampton, Frank A. *Deep Enough: A Working Stiff in the Western Mine Camps*. Denver, 1956.

A classic of the mining frontier, undeservedly neglected because of its use of mining slang and technical words. Crampton was a tramp miner in the Goldfield, Tonopah, and Great Basin rushes of the period after 1900.

De Kalb, Courtnay. "William Morris Stewart," *Mining and Scientific Press*. Vol. XCVIII (May 1, 1909), 599–600.

Egenhoff, Elizabeth L. "The Cornish Pump," California Division of Mines and Geology *Mineral Information Service*, Vol. XX (June, August, 1967), 59–70, 91–97.

[E. I. DuPont de Nemours & Co.]. *Blaster's Handbook*. Wilmington, 1958.

Foster, C. Le Neve. *A Text-Book of Ore and Stone Mining*. London, 1897.

This review deals with the pure Cornish mining canon at full flower. Among the illustrations are pictures of Cornish women in local costume sorting and cobbing tin ore.

Knox, Thomas W. *Underground; or, Life Below the Surface*. Hartford, 1876.

This book should be regarded as an item of curiosa, being evidently a newspaperman's paste-pot compilation of everything below the earth from the sewers of Paris to the silver lodes of Tucson.

Lewis, Robert S. *Elements of Mining*. 3d rev. ed. by George B. Clark. New York, Sydney, and London, 1964.

McCarthy, E. T. *Incidents in the Life of a Mining Engineer*. New York and London, 1918.

The chief interest of this autobiography is the author's experience in the lodes of the southern Appalachians in the Reconstruction period.

Paul, Wolfgang. *Mining Lore*. Portland, Oreg., 1969.

A compilation of things interesting or ornamental concerning mining, dealing chiefly with Germans and Germany.

Rickard, Thomas A. *Retrospect: An Autobiography.* New York and London, 1937.

To paraphrase Mr. Dooley, these are the words of the hero as they fell from his own lips and were taken down by his own hand.

Sandström, Gösta. *Tunnels.* New York, Chicago, and San Francisco, 1963.

A ponderous, well-illustrated history of tunneling from ancient times to the present day, by an engineer-journalist specializing in the field.

Simonin, Louis. *Mines and Miners; or, Underground Life.* Trans. and ed. by H. W. Bristow. London, 1869.

This remarkably thorough and well-illustrated work appeared in 1867 as *La vie souterrane.*

Stoces, Bohuslav. *Introduction to Mining.* 2 vols. London, 1954.

Taylor, Edgar. "High Explosives and Safety-Fuse," *Mining and Scientific Press,* Vol. XCVIII (May 22, 1909), 726.

Young, George J. *Elements of Mining.* New York and London, 1946.

Young, Otis E, Jr. "Thomas Arthur Rickard, 1864–1953," *Arizona and the West,* Vol. XI (Summer, 1969), 105–108.

Ancient and Renaissance Sources

Agricola, Georgius [Georg Bauer]. *De re metallica.* Trans. and ed. by Herbert Clark Hoover and Lou Henry Hoover. New York, 1950.

First published in 1555, this is the first and greatest of all textbooks on mining, translated and edited by a great mining engineer. Unfortunately, Bauer felt constrained to write in a contrived Latin, with the result that many of his terms are obscure. Also unfortunately, the English translation lacks a glossary.

Aristotle. *Works.* Ed. by W. D. Ross. 12 vols. Oxford, 1963–67.

Biringuccio, Vannoccio. *The Pirotechnia of Vannoccio Biringuccio.* Trans. and ed. by Cyril S. Smith and Martha T. Gnudi. Cambridge, Mass., and London, 1959.

Craig, Sir John. *The Mint: A History of the London Mint from A.D. 287 to 1948.* Cambridge, England, 1953.

Diodorus Siculus. *Diodorus of Sicily.* Trans. by C. H. Oldfather. 12 vols. Cambridge, Mass., and London, 1960.

Ercker, Lazarus. *Treatise on Ores and Assaying.* Trans. and ed. by Anneliese Grünhaldt Sisco and Cyril Stanley Smith. Chicago, 1951.

Herodotus. *The Histories.* Trans. by A. D. Godley. New York and London, 1957–61.

Pliny. *Natural History*. Trans. by H. Rackham, W. H. S. Jones, and D. E. Eichholz. 10 vols. Cambridge, Mass., and London, 1938–63.

El sistema del rato

Bailey, L. R. *Indian Slave Trade in the Southwest: A Study of Slave-taking and the Traffic in Indian Captives*. Los Angeles, 1966.

Bernstein, Marvin D. *The Mexican Mining Industry, 1890–1905*. Albany, 1964.

Catlin, George. *North American Indians*. 2 vols. London, 1866.

Dobie, J. Frank. *Apache Gold and Yaqui Silver*. Boston, 1939.

———. *Coronado's Children: Lost Mines and Buried Treasure of the Southwest*. Dallas, 1930.

Emmerich, André. *Sweat of the Sun and Tears of the Moon: Gold and Silver in Pre-Columbian Art*. Seattle, 1965.

García, Trinidad. *Los mineros Mexicanos*. Mexico City, 1895.

Gilliam, Albert M. *Travels over the Table Lands and Cordilleras of Mexico*. Philadelphia and London, 1846.

Howe, Walter. *The Mining Guild of New Spain and Its Tribunal General, 1770–1821*. New York, 1968.

Humboldt, Alexander von. *Political Essay on the Kingdom of New Spain*. Facsimile of 1811 ed. 3 vols. New York, 1966.

Johnson, Kenneth M. *The New Almaden Quicksilver Mine*. Georgetown, Calif., 1963.

Köhnicke, Theodoro. "Inexpensive Home-made 20-Ton Mill," *Mining and Scientific Press*, Vol. XCVII (August 8, 1908), 185–86.

Manje, Juan Mateo. *Luz de tierra incógnita*. Trans. by Harry J. Karnes. Tucson, 1954.

Mannix, J. Bernard. *Mines and Their Story*. London, 1913.

Merrill, F. J. H. "Dry Placers of Northern Sonora," *Mining and Scientific Press*, Vol. XCVII (September 12, 1908), 360–61.

Motten, Clement J. *Mexican Silver and the Enlightenment*. Philadelphia, 1950.

Probert, Alan. "Bartolomé de Medina: The Patio Process and the Sixteenth Century Silver Crisis," *Journal of the West*, Vol. VIII (January, 1969), 90–124.

Rickard, Thomas A. *Journeys of Observation* [including *Across the San Juans*]. San Francisco, 1907.

Ward, Sir Henry E. *Mexico in 1827*. 2 vols. London, 1828.

Wilson, Irma. *Mexico: A Century of Educational Thought*. New York, 1941.

Zavala, Silvio. *Los esclavos Indios en Nueva Espana*. Mexico City, 1967.

Placering

[Arizona Bureau of Mines] *Gold Placers and Gold Placering in Arizona.* Tucson, 1961.

Boericke, William F. *Prospecting and Operating Small Gold Placers.* New York and London, 1933.

Burgess, Hubert. "Anecdotes of the Mines," *Century Magazine,* Vol. XLII (1891), 269–70

Caughey, John W. *Gold Is the Cornerstone.* Berkeley and Los Angeles, 1948.

Gudde, Erwin G. *Bigler's Chronicle of the West: The Conquest of California, Discovery of Gold, and Mormon Settlement as Reflected in William Bigler's Diaries.* Berkeley and Los Angeles, 1962.

Haley, Charles S. *Gold Placers of California.* San Francisco, 1923.

Kelley, Robert L. *Gold vs. Grain: The Hydraulic Mining Controversy in California's Sacramento Valley, a Chapter in the Decline of the Concept of Laissez Faire.* Glendale, 1959.

Paul, Rodman W. *California Gold: The Beginning of Mining in the Far West.* Lincoln, 1947.

Rickard, Thomas A. "Farmer v. Miner," *Mining and Scientific Press,* Vol. XCVIII (April 24, 1909), 566–67.

Shinn, Charles H. *Mining Camps: A Study in American Frontier Government.* New York, 1948.

A most valuable study of the development of mining common law and *ad hoc* adaptation of the Spanish alcalde system to gold-rush California needs.

Tyler, Daniel. *A Concise History of the Mormon Battalion in the Mexican War, 1846–1847.* Chicago, 1964.

Waite, E. G. "Pioneer Mining in California," *Century Magazine,* Vol. XLII (1891), 132–40.

Weatherbe, D'Arcy. *Dredging for Gold in California.* San Francisco, 1907.

Prospecting and Assaying

Albert, Herman W. *Odyssey of a Desert Prospector.* Norman, 1967.

Reminiscences of a Nevada prospector contemporary with Crampton.

Anderson, J. W. *The Prospector's Handbook: A Guide for the Prospector and Traveler in Search of Metal-bearing or Other Valuable Minerals.* London, 1921.

Ansted, David T. *The Gold-Seeker's Manual.* London, 1849.

A very scarce early account of placering as it was moving from pan and rocker into more complex techniques.

Bernewitz, Max v. *Handbook for Prospectors and Operators of Small Mines.* Rev. and ed. by Harry C. Chellson. New York and London, 1943.

The most modern, comprehensive, and authoritative of the prospector's handbooks.

Bugbee, Edward E. *A Textbook of Fire Assaying*. New York and London, 1933.

Frazier, Samuel M. *Secrets of the Rocks; or, the Story of the Hills and Gulches: A Manual of Hints and Helps for the Prospector and Miner.* Denver, 1905.

Excellent on blowpipe field assaying.

Johnson, Joseph C. F. *Getting Gold: A Practical Treatise for Prospectors, Miners, and Students.* London, 1917.

Prospecting methods and tricks as used in Australia, a region American historians generally overlook.

Lakes, Arthur. *Prospecting for Gold and Silver in North America.* 2d ed. Scranton, 1896.

Lakes was a famed paleontologist and a professor at the Colorado School of Mines. The historical information on Colorado is excellent. Lakes misunderstood much of the genesis of ore, but his book gives insight into what the geologists and promoters of his period thought about the subject and can be checked against Emmons (see above) and others to avoid substantive error.

Miller, G. W. *The Mine Examiner and Prospector's Companion.* Denver, 1903.

Morris, Henry Curtis. *Desert Gold and Total Prospecting*. Washington, D.C., 1955.

The fragmentary, unorganized, but very interesting recollections of a young mining engineer whose career began at Tonopah and Goldfield.

Young, Otis E, Jr. "The Craft of the Prospector," *Montana: the Magazine of Western History,* Vol. XX (January, 1970), 28–39.

———. "The Prospectors," in *Reflections of Western Historians.* Ed. by John A. Carroll. Tucson, 1969.

Mining Districts

Billeb, Emil E. *Mining Camp Days.* Berkeley, 1968.

Largely concerned with later days in the Death Valley and Mojave Desert gold camps in southeastern California.

Blake, William P. *Tombstone and Its Mines.* New York, 1902.

Though apparently a promotional brochure, perhaps written at the behest of Frank Murphey, this is one of the very few nonprofessional publications devoted to Tombstone as a mining camp.

Brown, Robert L. *Ghost Towns of the Colorado Rockies.* Caldwell, Idaho, 1968.

A ghost-town album interesting for its photographic views of former Colorado mining camps.

Browne, J. Ross. *Adventures in the Apache Country: A Tour Through Arizona and Sonora with Notes on the Silver Regions of Nevada.* New York, 1869.

Browne was a promoter and journalist of uncertain reliability, but his early arrival in the region makes him a primary source.

Burt, Silas W., and Edward L. Berthoud. *The Rocky Mountain Gold Regions, Containing Sketches of Its History, Geography, Botany, Geology, Mineralogy and Gold Mines, and Their Present Operation: With a Complete History of the Mills for Reducing Quartz Now Operating.* Denver, 1962.

A facsimile edition of one of the primary works on early Colorado.

Clemens, Samuel L. [Mark Twain]. *Roughing It.* Hartford, 1872.

Mark Twain's classic of Washoe flush times requires little comment except that he was a better reporter than he has sometimes been given credit for, and a worse miner than he gave himself credit for being.

Dunning, Charles H., and Edward H. Peplow, Jr. *Rock to Riches: the Story of American Mining.* Phoenix, 1959.

The title is a misnomer since the book deals exclusively with Arizona. It is badly organized and worse written, but its introductory chapter on mining is good, and it contains essential information on all phases of Arizona mining and placering.

Elliott, Russell R. *Nevada's Twentieth Century Mining Boom: Tonopah, Goldfield, Ely.* Reno, 1966.

A sound, well-documented, scholarly account of the end of the mining frontier in Nevada and the beginning of modern mining in the copper porphyries of Ely.

Harpending, Asbury. *The Great Diamond Hoax and Other Stirring Events in the Life of Asbury Harpending.* Ed. by James H. Wilkins. Norman, 1958.

Hinton, Richard J. *The Handbook to Arizona: Its Resources, History, Towns, Mines, Ruins and Scenery.* San Francisco, 1878.

Hollister, Ovando J. *The Mines of Colorado.* Springfield, Mass., 1867.

Perhaps the primary source for J. H. Gregory's discovery of the Central City lodes. The book is very scarce and should be reissued.

Hughes, Richard B. *Pioneer Years in the Black Hills.* Ed. by Agnes Wright Spring. Glendale, 1957.

Jackson, W. Turrentine. *Treasure Hill: Portrait of a Silver Mining Camp.* Tucson, 1963.

A sound account of the White Pine silver camp of the Great Basin, interesting for its investigation of British investment in American mining.

Kelly, J. Wells. *First Directory of Nevada Territory.* Facsimile ed. Los Gatos, 1962.

The source for detailed accounts of very early quartz milling on the Comstock Lode.

Lord, Eliot. *Comstock Mining and Miners.* Facsimile ed. Berkeley, 1959. (Original ed., 1883).

A reissue of one of the major sources on the Comstock Lode, with added illustrations of great value. Lord was not immune to error and was tender toward Ralston and Sharon. Even so, the original publication, sponsored by the United States Geological Survey, created something of a scandal.

Lyman, George D. *Ralston's Ring.* London, 1937.

A secondary account of the machinations of Ralston and Sharon on the Comstock Lode. The writing is in a popular vein, and the author does not attempt to conceal his dislike for both men.

Parker, Watson. *Gold in the Black Hills.* Norman, 1966.

Mineral-district history as it should be written, with close attention to the sources and a critical approach to the entire subject. It should serve as a model for the other badly needed histories of such districts. Virtually every district is celebrated in some book, but too many of them are undocumented, fail to include bibliographies, invent conversation, repeat discredited legends, and bear colophons of vanity publishers.

Potter, Alvina N. *The Many Lives of the Lynx.* Prescott, 1964.

Despite the absence of documentation and bibliography, this amateur history of the Bradshaw (Lynx Creek) mining district of Arizona is redeemed by the author's evident industry and sincerity.

Rickard, Thomas A. "Rich Ore and Its Moral Effects," *Mining and Scientific Press,* Vol. XCVI (1908), 774–75.

Reflections upon high-grading and related vices at Goldfield.

[Schieffelin, Edward]. "History of the Discovery of Tombstone, Arizona, as Told by the Discoverer, Edward Schieffelin." Typescript in the archives of the Arizona Pioneers' Historical Society, Tucson. (Original manuscript in the Bancroft Library, University of California, Berkeley.)

This memoir represents one of the few primary sources on the Tombstone strike.

Shinn, Charles H. *The Story of the Mine, as Illustrated by the Great Comstock Lode of Nevada.* New York, 1896.

One of the three great works on the Comstock Lode, the others being those by contemporaries Lord and Wright.

Smith, Grant H. *The History of the Comstock Lode, 1850–1920.* University of Nevada *Bulletin*, Vol. XXXVII, No. 3 (July 1, 1943).

Smith's scholarly investigation of the Comstock Lode sets right many of the errors perpetrated on or by Lord, Shinn, and Wright. Despite its title, the book has nothing to say about the cyanide period. The author's statistics on investment and production are important to an understanding of the American mining frontier.

Stewart, R. E., and M. F. Stewart. *Adolph Sutro.* Berkeley, 1962.

An adequate biography of Sutro and his tunnel.

Symms, Whitman. "Decline and Revival of Comstock Mining," *Mining and Scientific Press*, Vol. XCVII (October 12, 24, 1908), 496–500, 570–76.

Walker, Elton W. "Sinking a Wet Shaft at Tombstone," *Mining and Scientific Press*, Vol. XCVIII (February 20, 1909), 284–86.

Wilson, E. D., J. B. Cunningham, and G. M. Butler. "Arizona Lode Gold Mines and Gold Mining," *University of Arizona Bulletin* Vol. V, No. 6, (August 15, 1934); Arizona Bureau of Mines, Mineral Technology Series No. 37, Bulletin No. 137.

This monograph contains a detailed register of the innumerable small Arizona gold lodes. The concluding section is interesting as a review of small-scale gold mining.

Wilson, Neill C. *Silver Stampede: The Career of Death Valley's Hell-Camp, Old Panamint.* New York, 1937.

A characteristic example of the popularized mining-camp or mineral-district history.

Woodard, Bruce A. *Diamonds in the Salt.* Boulder, 1967.

A well-documented account of the Great Diamond Hoax and its perpetrators, Arnold and Slack. Both this work and Clarence W. King's original United States Geological Survey report ought to dispose of the persistent claim that the site of the hoax was Arizona.

Wright, William [Dan deQuille]. *The Big Bonanza: An Authentic Account of the Discovery, History, and Working of the World-renowned Comstock Lode of Nevada.* New York, 1964.

Wright, a reporter for the *Territorial Enterprise*, was accorded liberties by Washoe mine superintendents which gave him access to information denied others. Nonetheless, on the evidence, Wright appears to have had no training in any aspect of mining or milling and to have absorbed nothing more than what he was told in conversation.

Index